HOME ON THE WAVES

A Pacific Sailing Adventure

by Patrick Hill

PROMONTORY PRESS

Home on the Waves
Copyright 2015 by Patrick Hill

All rights reserved. No part of this book may be used or reproduced in any matter without prior written permission.

Promontory Press
www.promontorypress.com

Cover design by Marla Thompson of Edge of Water Designs

ISBN: 978-1-927559-94-9

Printed in India

My Thanks

My thanks go to my wife, Heather, whose positive thoughts and unstinting support made the whole dream a wonderful reality and to Jeremy and Erica, our children, whose enthusiasm and reactions to the many adventures were a delight to experience.

As this story is being written I thank my family again for their support and their input and recall of so many happenings that have been cloaked by time.

Also, my thanks to our good friend, Joanna Gould, who provided format and text improvements and supporting thoughts.

Foreword

While trekking recently in the vastness of Nepal, we were treated to breathtaking scenes of huge snow-covered peaks, moss-covered forests, deep gorges, roaring rivers, barren open table lands, steep terraced hillsides, mule and yak caravans, tiny villages – some remote, some derelict looking, and a wonderful variety of friendly Nepalese and Tibetan people.

I was travelling with my wife, Heather, our guide, Chyangba and his porter, Guncha. As we trekked up and down the many steep trails, cliff-side paths with dizzying drops below and mountain ridges, we moved at our own pace leaving much time for my thoughts to range far and wide, uncluttered with the multitude of home life activities and only the wonder of nature around me. Scenes from the past kept whirling in and out of my mind from work, my early life, trips we had made, trips I would like to do and thus in this thought process, hopefully not the first sign of old age rambling, I decided I would write a book.

Books are perhaps man's widest diversion. They entertain, inform, instruct, philosophize, titillate and most cause one to think and dream. Also with the dreaming may come the realization that one can achieve objectives earlier thought not possible. This book is a simple story of one of the key events in my life with Heather, that of a sailing trip to the South Seas with our family. I hope it will provide a refreshing reminder for our children of a year's trip we had together. Hopefully too, it may help to motivate those who are poised but hesitant to explore life further. No words can express this better than those of the poem of our late young friend, Richard Ellott of Whistler, British Columbia:

> "As an eagle takes flight
> And is borne above the clouds
> So too must I spread my wings
> And embrace the power of the wind"

Patrick Hill, Kathmandu

For copies: www.patrickhillcruising.com

Contents

	My Thanks	iii
	Foreword	v
1	Planning	1
2	Our Floating Home	7
3	Last Thoughts On Leaving	26
4	Off to San Francisco	34
5	Harbour Hopping to San Diego	54
6	Mexico	88
7	Three Weeks at Sea	116
8	French Polynesia	137
9	Rough Passage to Hawaii	196
10	North Pacific to Glacier Bay	209
11	Homeward Bound	235
	Afterthoughts	245
	Appendix: Equipment	250
	Glossary	254

CHAPTER 1

Planning

"as a wind works its magic on the sails"

1970 . . . . From my office window I watched the yellowish silt-laden waters flow around Spanish Banks and ease their way into English Bay on the incoming tide. With this natural process the shallow Banks continued to grow outwards sometimes to the inconvenience and surprise of the unwary yachtie as he touched bottom. Dark silhouettes of freighters high in the water, waiting for berths at the harbour facilities, swung at anchor in no particular pattern as currents and winds produced conflicting influences. Faintly in the distance across the Strait of Georgia shadows of mountains loomed through the misty air. These magnificent views which enchanted and enticed me so many times were lessening with the continued growth of high-rise buildings as Vancouver discovered itself and emulated the big-city look. It emphasized my sense of confinement that had been growing over the past year, perhaps years.

I sat there musing and asked the very familiar questions that most of us experience from time to time, *"What am I doing and where am I going?"* It seemed some further challenge, some further fulfillment was needed. I should not just continue to sit on my comfortable "pot". Yes, the "pot" was pretty nice. Here I was, age 38, happily married to Heather, and with two children: Jeremy (9) and Erica (5). That dreadful millstone, the mortgage, all $8,500 of it, was more or less under control, and we could see that day when the loan agreement could be appropriately destroyed while consuming at least a moderately priced champagne. I was a professional engineer with a major international engineering company and was fortunate to experience challenging and variable work while travelling extensively. Still, there were so many other things to do and see besides work. Those who advocate play now and work later do have a point.

Sailing was our passion so the oceans beckoned. Gliding through calm waters as a wind works its magic on the sails, dropping anchor in lonely bays, just getting the maximum speed out of a boat regardless of the wind, or just lolling around with family and friends – what better way to go?

We had experienced two boats. A centre-board 16' Snipe which we had kept anchored off MacDonald's yard next to the Kitsilano Yacht Club years ago. We sold her when we went to Australia. Our next boat came four years later back in Vancouver. We were then a family of four and reckoned we needed about a 26' boat. Most fibreglass ones were out of our range, unless I dropped my old fashioned ideas of paying outright in cash. Finally we zeroed in on the ubiquitous 26' Thunderbird going at a reasonable price and bought one in partnership with a friend who had a similar shortage of cash.

I have never been a great one for nautical terms but, when

Planning

I went to view an earlier Thunderbird, the owner insisted on taking me out for a sail, believing – I think – that I would be a quick buyer. He threw me two bundles of ropes saying that I could put on the "sheets" while he got the genny hanked on. Not wishing to expose my ignorance as to what a sheet was, I spent the longest amount of time blowing my nose until he picked the bundles up and attached them to the genny. Then I knew what sheets were!

This somewhat tender boat, designed for the lighter, average winds of the sheltered areas of the west coast, provided some marvellous and exciting sailing with its extra expanse of sail. It was a fine and sleek looking boat in the water and out, with its highly efficient fin keel, clean underbody and large cockpit. It was quite influential on the kind of boat we were looking for. I managed, with some adjustment, to get a 10 HP outboard long shaft motor, usually hung on the transom, to fit inside the transom lazerette; by rotating it out of the water and within the line of the hull, a cover plate could be installed to make a completely flush hull when sailing. It was a fine buy, certainly compared to some production boats, one of which actually had an outboard motor installation right in the middle of the cockpit!

"saying goodbye to a very inbred work ethic"

A year later, we bought the other half from our friend and kept the boat, *Kolus*, for another three years of fun sailing holidays in the warm waters of Desolation Sound and the Gulf Islands. Destinations at Plumper Cove, Keats Island or any of the four bays on the south side of Gambier Island, were convenient weekend sails, all this within 10 miles of our city berth.

On Friday nights, Heather, equally enthusiastic about sailing, would meet me at my office with the kids and all the goodies required for a weekend. While Heather drove to the boat I would change out of my office gear and we'd be gone early and back late Sunday night. By 6pm we would be headed up the coast with work forgotten and looking for a place to anchor.

This very affordable boat had provided us with several years of wonderful memories, of super holidays, moments of exhilaration and moments of concern. The day the cat jumped ship, we believe in preference for an adjacent fish boat, was one of mourning. The coach roof corners of these boats frequently leaked where the kids were sleeping; our short term remedies were to take out the thin wood panels from under two seats and use them to deflect the drips to the side of the boat. Use of the toilet required a certain amount of tact from the rest of the crew.

Sometimes things would get out of hand such as when we first raised the spinnaker, or if the children got too much, we would tow them 100' behind in the dinghy. If things got too much for Heather, she would be towed. The children developed a new submarine technique that I am sure Sportyak never had in mind for their dinghies; as we towed them, they would flood it by moving up into the bow and then with skillful adjustment of their weight, plane the dinghy downwards until only their heads showed above the water or, sometimes, they went right out of sight.

Our west coast is blessed with the most superb scenery, islands, and sheltered waters; they stretch for 900 miles to Juneau, a unique coastline. Our trips would range up to 100 miles to areas around Desolation Sound (no doubt named by Captain Vancouver in his search for the Northwest Passage). Now however, my longing, and, I believed, Heather's too, was to explore further afield. Warmer waters and cruising beyond the normal holiday areas beckoned. We certainly did not have the money to buy a boat that could take us safely away and offshore. It seemed that we would have to build one. This would certainly meet some of my desire for a challenge and some creative work of a different nature.

Back in England where we immigrated from, the idea of owning a sizable boat was totally out of the question. That kind of money was not available in my family, but perhaps more than that, the concept of owning a boat did not occur to one any more than owning a Jaguar or a Bentley. As an engineering student, my budget was just minimal, as were the ten cigarettes I allowed myself for the week. Sometimes they were almost all consumed on a single night out, trying desperately to impress some damsel.

Building a sailboat would require a big effort. Could I do this? It was not only the building of something completely foreign to me but it was the concept of later saying goodbye to a very inbred work ethic as well as possible problems of later returning to that environment. So, step one was to see if the family was positive and think about building.

Thin lines of silver grey spread across the waters of English Bay in the dying daylight as I left the office for home……………………

Our family discussions were spread over many weeks as the various aspects were considered. Good friends were consulted as we thought about what to build, costs to build, costs to go sailing, education for Jeremy and Erica, what to do with our house and especially the reluctance from one member of the family whose preference was for horses. Boating books and magazines were consumed endlessly as marinas were searched and researched.

We wanted a fibreglass boat and wanted nothing to do with wood, concrete, steel or aluminum materials for various specific reasons. Naval architects were consulted but their concepts and high prices for plans did not excite us. At the time, we were looking for broad-beamed boats with the beam carried well aft, to give volume and stability, and with moderate displacement; these did not seem to exist. One European yachting magazine referred to the "... enormous 11' beam..." on a 60' yacht. This was not our idea which was at least 12' beam on a 40' boat. In any case, one quote for plans of a boat with our requirements was nearly a quarter of my proposed building budget!

Eventually, a recommendation came from Charlie Kennedy, a good South African friend who had built boats and sailed offshore. He suggested that the jigs of a 42' boat (a Fraser 42) which were for sale, would form a good basis from which to start building. It would suit our needs for future offshore cruising! The jigs belonged to Jim Innes, who, along with a number of other airline pilots in Vancouver, had put together this design specifically to race to Maui. The jigs, made of wood, provided the hull cross-section shape at 4' centres through the boat.

One day, Heather and I turned up at his house and were swept up with the enthusiasm of Jim and his wife, Shirley. Jim was a dynamic character who had by then built 2 or 3 houses, a small 'plane and the yacht *Long Gone* from the jigs. While Jim had built his Fraser 42 from jigs as a "one-off", the other pilots had proceeded with production female molds for hull and deck from which to produce their boats. For us to purchase a finished hull and deck would have been too expensive. We had even approached the owners of the molds to see if we could rent them to fibreglass our own hull and deck but they were not receptive.

However, with Jim's descriptions of the performance of his boat and viewing of his sailing film we were sold. We viewed this boat as the answer to our long search and, without hesitation, happily plonked our money down ($500) and had the jigs delivered to our house. We were on our way.

CHAPTER 2

Our Floating Home

"Er!Er! Mr Hill, can you tell me how you got an industrial zoning for your property?"

So we built this centre-cockpit 42' boat from Jim's jigs. It was a fascinating process which invaded our lifestyle. Up to this point, the biggest thing I had built, excluding work projects, was a balcony. I had taken a short evening course on fibreglassing in preparation for this project. The construction was of fibreglass with an Airex foam core sandwich hull and a balsa cored deck. We made many significant modifications to the original design, including installing a taller mast; increasing the keel weight; making the keel shorter and thinner; adding a skeg to make the boat track better; and changing the conventional transom to a reverse transom, which not only looked better but gave an extra foot of space inside. We also changed the deck layout to give more room below.

She grew in our small garden from six drums of resin to a sleek beauty. It took us 6,000 hours over a three-year period to build. Jeremy, who works in the marine industry, kindly advised that he has never met or heard of anyone completing a boat from scratch in that time. Since I was working about 2,000 hours a year in my normal job, the extent of this task can be appreciated. It was a huge amount of work accomplished with tremendous support from Heather.

It is extraordinary where energy comes from when one is really motivated. One of the fastest and simplest restorers of energy for me was the catnap. Sometimes on a hot Sunday afternoon when the adrenaline flow was slowing, I would lie down on the cabin floor amid the mess and just "drop off". Five or ten minutes later, a sudden snort would bring me back to reality rejuvenated to full speed; a marvellous body function which Heather was not entirely romantic about!

Sometimes Heather and I would do some fibreglassing before I left for work. When I returned and caught a whiff of fibreglass at the end of our cul-de-sac, the juices would start to flow. When the kids had gone to bed, I would investigate the changes of plans and layouts required, as well as the items needing purchase. Heather made cushions, became a good fibreglasser, and was a ready helper; just seeing my normal working mess cleaned and swept up was always a big lift. Chris, her dad, would sometimes help as would other friends from time to time. Visitors with positive thoughts and help were welcomed, others with negative thoughts were avoided. I didn't want to hear comments such as, "*You have so much to do, will you ever finish it?*" People with these thoughts were diplomatically eased away by Heather. Other boat builders who visited were welcomed, particularly Charlie Kennedy who would offer good advice or say, "*Just wait till you feel her riding over the waves*". Sometimes he would be sitting in his car just ready to leave and he might say, "*You know, Patrick, one more glazing of the hull would make it really look good"* and I would inwardly sigh knowing he was right but that it would take me another month of work.

Needing some boat knowledge for the changes I was proposing, I was in a bookshop during my lunch hour looking through an interesting design book when a voice behind said, *"That looks like a useful book."* Roy Hamilton, also an engineer, told me he was building a smaller boat. I said, *"Look I am going to buy this so why don't you buy it with me and we can share it as we build?"* He immediately agreed and we maintained a great relationship thereafter.

However, the building is the subject of another book, but there were a couple of exciting and tense times during the project. Our excitement rose as the crane operator lifted the finished hull from its supports in the back garden, rotated it from its upside down position and we saw, for the first time, what this boat really looked like, Holy Smoke, at 42' long, it was big! The operator continued the lift and swung the hull over the top of our house onto prepared supports at the start of our driveway. Phew! This very emotional moment after six months work was followed by the first of many building parties.

Another tense moment was the pouring of the 4-ton lead keel. I had split a 45-gallon drum down the middle, and in each of the two halves I put 4,000 lbs of lead pellets. The drum halves were sitting on a stand in our small front garden. Between the halves I had packed 200 lbs of coke. Very early on this critical day, I lit the coke and used our long-suffering vacuum cleaner to blow air through a steel tube into the coke to increase the heat. Then I started four propane torches on the outsides of the halves. It was quite a scene. Later our good-humoured elderly neighbour asked over the fence in a querulous voice, *"Er! Er! Mr Hill, can you tell me how you got an industrial zoning for your property?"*

Four hours later, specially selected friends including a doctor, a fireman, a carpenter, two engineers, and a labour specialist, arrived to help. Jeremy and his school friend, Dennis (the fireman's son), took the day off from school to watch. The moment of truth arrived as I broke the two seals and watched two jets of molten lead stream into the concrete keel mold I had made. I felt like some medieval alchemist. Suddenly, I realized that we might be 400lbs short due to the waste slag in the pellets, so Heather was immediately dispatched to the local metal yard to get four 100lb lead ingots which we melted into the mold at the end. Heather complained that with the weight right at the back of our old Rambler station wagon it tipped so much she could hardly see over the wheel to drive back!

Throughout, the building seemed to be of considerable interest to our neighbours who would visit, perhaps really wanting to know when the damn thing would be finished; I did have them signed up to allow me to build for two years. The kids were always interested and often had sleep-overs on board in the later stages. At the completion of each significant stage we had a party with friends, neighbours, and other boat builders; we had quite a few parties. At the last party when building was finished, I gave back to all our tolerant neighbours wrapped presents of all the tools they had lent us.

Later, when the winds thrust her forward, memories of the long hours and months of labour faded to be replaced by a great feeling of achievement and expectation. Watching this "baby" being towed out of our driveway for its trip down to Mosquito Creek was another highly emotional moment for us (seemingly also for our neighbours – now they could sleep in on Sundays) as was its champagne launching filmed by our local North Shore News. Earlier, after much soul searching, we named her *Sky One Hundred*, derived from Erica's expression for something big: *"I love you Mummy one hundred and*

Our Floating Home

the sky".

At 18,500 pounds, our floating home was of moderate displacement with the layout shown in the picture. There was a large forward cabin with a full vee-berth and cupboard space. Heather and I slept there. I liked to sleep forward as I could hear the movement of the anchor, sometimes necessary on a really windy night. Heather always insisted it did not matter where I bunked down as I always slept like a log.

Directly aft was the main cabin which consisted of a 7' berth on the starboard side built into the curve of the hull. There was always a rush for this berth-come-couch, it being the most comfortable seat. Opposite there was a u-shaped dining area with a table. On both sides of the cabin the seats had shelves above and storage lockers below. Above the dinette area there was a wine cupboard.

Aft of the dining area was the 4'x4' galley; although small it was the key action area. It had one sink, a counter with a well insulated ice box below, and a cupboard above. Located almost on the boat centerline was our propane stove with an oven and three burners. Unlike most production boats the stove was not gimballed. It was also located under the cockpit seat where the cook of the moment could not be thrown onto it by rough seas, as might happen if it were located in the open by the hull. A waist-level strap could hold the cook in position on rough days so that both hands could be used for cooking.

Aft again of the long berth was our large head. We built it there as having the least boat motion at sea (perhaps I should say least "movement"); a necessary requirement for the convenience of one's daily constitutional. We didn't want the head located forward, as in most boats, where two hands might be required to ensure you didn't actually part company with the "seat" on a rough day. Just forward of the head we had a simple pull-down board for any chart work.

Companionway steps led out of the main cabin hatch into a large cockpit with seats on three sides except across the companionway; here we had left the seat out so that we did not have to keep stepping over it when using the steps. Under the steps the engine was located in a simple break-apart box so that it could be fully exposed and easy to work on. Close to the top of the steps was a locker under the cockpit seat for two large propane tanks which could be manually turned on and off easily, as required, from the bottom of the companionway steps. The locker had an outboard drain to allow any possible heavier-than-air propane gas to pass overboard and not into the boat bilge where it potentially would present a major hazard. As a precaution, I would regularly check with my nose at floor level for smell of any propane in the bilge – there was no automatic sniffing detection device.

The cockpit had wheel steering at its aft end where a hatch led into the aft cabin. This cabin could also be accessed by a crouch through-way from the galley; on one side of this were shelves for food storage and on the other, a sliding door to a small engine room to access the transmission, stuffing gland, and other minor equipment. In the aft cabin there was a large double berth right in the stern. Half of this berth was used for storage of school books and Jeremy's and Erica's gear; the other half was used for Jeremy's bunk. A large quarter berth for Erica existed on the starboard side of the cabin.

This was a functional keep-it-simple type of boat. The only electronics were in the ham radio set, the VHF radio (for in-sight communication), the depth sounder, and the log (for boat speed). All other operations were manual, such as the pumps for fresh and salt water; straight lengths of the ubiquitous metal coat hanger were used for checking the amount of fuel and water in the tanks; and, lowering and raising the anchor by hand (a straight pull of 120' of chain with a 35lb anchor was an effort once). This way little could go wrong.

This was to be our home for over a year – no 2,000 square foot house or large apartment. While it may seem such a small area the benefit was the minimal amount of movement required, saving of energy, and the ability to hang onto something close in rough weather. Heather and I had made up trial plywood and cardboard formats for each part of the boat to check and optimize the available space for each activity prior to final building. Now we were about to sample our choices.

Construction after one month

Launching day after 3 years

Our Floating Home

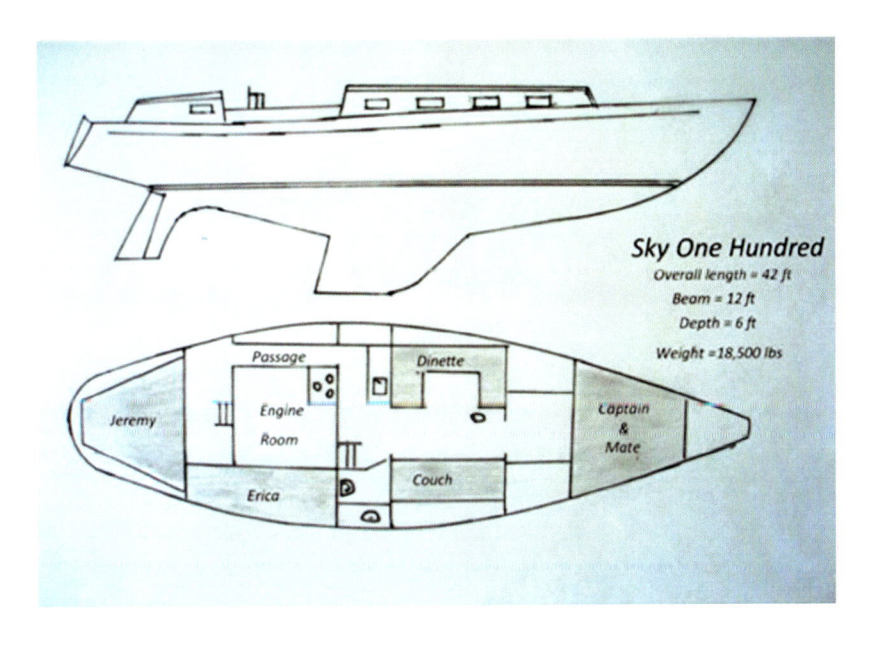

Sky One Hundred

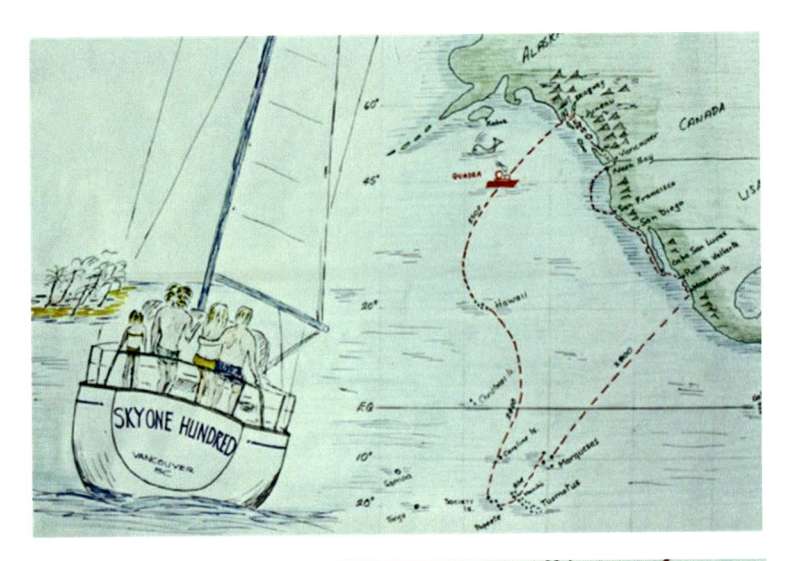

Our 15,000 mile trip

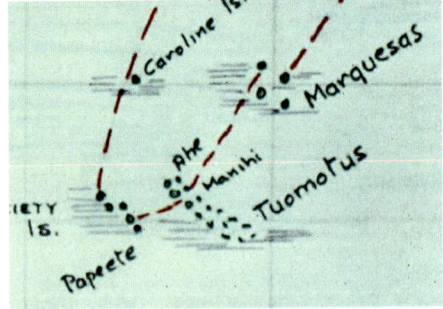

Big waves off the Oregon coast with Goldfinger working overtime

Our Floating Home

Bye bye San Francisco

Loading up for Mexico

Jeremy

19

Home on the Waves

CABO SAN LUCAS

Jerry, Heather and Randi
Erica and Renee

Shark fishing beach

Our Floating Home

Flying fish

Another
flying fish

Sunrise on the way to the
Marquesas

21

Home on the Waves

NUKU HIVA, MARQUESAS

BAIE DE TAIOHAE

Friendly locals

Simeon and Felicity

Waiting for church

Tahuata surf

Oa Pu

Camping gang

CHAPTER 3

Last Thoughts On Leaving

"So how often does one get some 12 years of holiday time on credit...?"

16 July '77 . . . . . .The rain was bucketing down. A solid grey blanket of cloud overhead offered no respite. This was the last thing we wanted on leaving Vancouver bound for the South Seas. However, it was the kind of image I projected when talking about Vancouver; I always hoped, selfishly, it would keep migration here to a minimum. More friends hailed us from the slippery dock at Mosquito Creek as we hauled the rain cover over the boom with sighs of relief from those in the cockpit. Down below the talk grew louder and the atmosphere warm and humid.

"Hi there, how come you picked a day like this?" "Do you guys really know what you are getting into?" "Here is something that will cure your seasickness" – another bottle! Oh good friends, we are certainly going to miss you all. More well-wishers climbed on the boat, another bottle – did they think we drink that much? A food package with a large tin of English biscuits, cans of fondue, and hard-tack biscuits – super! Another parcel, soft and cold – Good Heavens! It was a fully prepared 5-pound B.C. salmon. Well, I didn't think we would suffer too much in the first few days. A groan from below as the boat rolled disconcertingly in the bow wave of the newly installed Burrard Inlet ferry, the catamaran design "they" said would never produce a wave at 12 knots. I hoped the concept of our forthcoming trip would be closer to reality.

Did we know what we were really getting into? Were we fully prepared? Did we need additional crew?

We had asked ourselves these questions too many times. Now it was too late, we were committed. I had obtained a leave of absence for a year and we were now ready to go. I had handed over my various engineering projects. I did regret handing over a major design-construction project of which I was manager, but the waters beckoned; even my client could accept my leaving. Earlier I had invited my boss out to lunch during which we discussed the various engineering studies and projects underway. Towards the end he asked, "*Well, what's the pitch?*" I said I would like some extra time off and he said, "*Sure, take a week off.*" I said I needed more, to which he offered up another week or two provided projects would not suffer. When I said I needed more, he asked, "*How much do you want?*" I said I really needed a lot more – I wanted a year's leave of absence and explained my plans. I do believe a gleam of envy came into his eyes, he also being a boater; some time later he agreed. So how often does one get some 12 years of holiday time on credit . . . ?

We had planned a loosely scheduled trip departing in July and returning August the following year. Our first stop was to be San Francisco and thence coast-hopping down California to San Diego, over the Mexican border and down to Manzanillo, perhaps out to the Galapagos (a bit far south), the Marquesas Islands, the Tuamotu atolls (depending on how our navigation skills developed), Tahiti, and returning via Hawaii, the glaciers of Alaska, down the British Columbia inland passage, and home.

It was a pretty ambitious program involving at least 16 weeks of non-stop sailing over six ocean passages. However, there were some short cuts that could be adopted, such as omitting the Galapagos, Tuamotus, and Alaska. We had lots of time yet to adjust our circular tour. The main need was that our progress avoided poor seasonal weather conditions. Previous study of the Pilot Charts, Ocean Passages of the World, and particularly discussions with people who had made trips in these areas before, indicated the following: do not leave Vancouver before the end of April/May; pick up the prevailing north west winds; do not leave San Diego before the beginning of November due to the possibility of cyclones and the odd hurricane off the Mexican coast; pick up the NE trades some 200 miles offshore; do not go west of Tahiti until after March, Tahiti being the eastern limit of hurricanes; make as much easting as possible on the way to Hawaii; and, do not leave Hawaii until the beginning of June. They were the general guidelines and they fitted our trip pretty well.

I felt *Sky One Hundred* was a strong boat. During her building from scratch I had put extra fibreglass in key stress areas, such as the keel, chain-plate connections, the stem, transom, hull and deck connection, and the skeg. These were areas that often had problems in production boats with leaks or breakages. The rudder post was a 2.75" diameter stainless-steel shaft and the lead keel was attached with sixteen 1"diameter stainless-steel bolts through to its bottom. The mast was sturdy compared to some of the skinny types on racing boats. On a number of occasions on our trip we saw boats with these types of problems. One production wood boat (the type known as "leaky-teaky") leaked through the hull-deck connection all the way from Seattle to Hawaii. The owner said he had water running down the inside of the hull into the bilge. Particularly, I knew intimately each part of our boat.

We had raced it with some successes, ironing out the bugs and getting familiar with it and ourselves. We had raced her in the overnight 140-mile Georgia Strait race in which many of the boats did not finish and some lost their masts. She gave us no concern and actually tipped 14.5 knots (28 kms/hour) in a downwind surf. It was a rough race and I was sick for the first time. Heather was not sick; Heather is never sick! A good friend, Paul, who was becoming gung-ho on sailing and crewing for us was wretchedly ill overnight; on coming up on deck in the morning he took one look around, turned green again and went back to his prostrate position. His wife thanked us for taking him, saying it saved them from the purchase of a new yacht!

It is our belief that those contemplating offshore sailing should experience racing, even if they come near last each race! This forces the need to sail at night or beat to weather in unfavourable conditions; these situations are normally avoided when just cruising but do cause one to learn about the boat and in particular about one's self. Certainly one can learn how to make the boat go faster. We had heard tales of those going offshore without this experience and turning back. Sailing at night in nasty seas is not pleasant and perhaps many people do not like the loss of control.

The boat was fairly well prepared and we had been making modifications steadily for the last few months prior to our departure. I also expected to buy further additional equipment as necessary, further down the coast, with San Diego being our last main chance to stock up. Our main equipment is listed in the Appendix. We obtained correspondence courses for both Jeremy (Grade 11) and Erica (Grade 7) from the B.C. Department of Education. The children's schools were very helpful in providing required textbooks, plus extra ones they thought might be useful, and were enthusiastic about the trip.

I believed we had a well built boat but what of the crew? I did have some worries about the whole idea of thrusting my family off into oceans of which I had no experience. Strangely, the thought of ocean seas and being out of sight of land did not seem to concern me as much as being prostrated to a point of being useless with seasickness. Heather, while a Power Squadron member, was still mastering the navigation bit, as I was, and if I was incapacitated, well . . . ? We had had a few lessons from a friend and a bit of practice taking sights from the local beach much to the interest of passersby! Our friendly teacher would come to the house to instruct us and other off-shore dreamers. The trouble was that after a few drinks this single Norwegian would drift off to start telling us about *"the Heavenly bodies he had known . . . "* while we all groaned and had to get him back on course! I expected that Heather and I could get our navigation methods organized by the time we reached Neah Bay at the head of the Juan de Fuca Straits, before we hit the high seas.

While I might be on dangerous ground I thought Heather to be a capable sailor, and where I tend, at times, to get casual about passes or water depths, she was very thorough. Jeremy had done a lot of racing on our boat and others, and was very competent in that direction but was not so interested in navigation. Once *Sky One Hundred* was built and we had taken some family trips in it, he had wanted to take the boat out racing with his friends. I had said he could if he passed his Power Squadron course and got confirmation that my insurance would cover him which he did; so off he went in his first race with me looking at my diminishing fingernails! Erica was an unknown quantity because she was infinitely more interested in horses with a big "H" and would cheerfully have sold the boat to buy a ranch and all that went with it.

I thought we could manage without a crew – I certainly did not like the thought of other people around, friends or not. Friends leaving on a similar trip had had their own friends on their first leg to San Francisco but relations deteriorated to such an extent that they demanded to be put ashore on the Oregon coast. Unbelievably, they were taken off by a U.S. Coastguard cutter. As our moment of departure loomed, I still found myself concerned enough about the magnitude of this first step, with its usual rough seas, that I invited a young, experienced couple to join us on this leg. However, I was rather relieved when they declined; in spite of my concern I did want to complete this challenge without assistance.

Our most reliable crew was our wind vane which was used for steering the boat. Commercial vanes were very expensive and I had read of many failures of their sophisticated parts. I like to keep things simple so we made our own from an article by William Orgg in the May 1971 edition of *SAIL* magazine. I managed to simplify the design and built it for a tenth of the cost of a store-bought one. We called it "Goldfinger" because the vane was covered by Heather with some gold-coloured fabric purchased at Gold's Fabrics. "Goldfinger" was invaluable in that it consumed nothing and performed 24/7!

Most cruisers have to make a major investment in their boats, and off-shore insurance – an obvious must – was very expensive if not prohibitive. We managed to get a rider on our normal British Columbia waters coverage for a small charge, to cover us to San Diego. Having made this location safely we were able to extend this rider to Manzanillo in Mexico for a mere $25. From Mexico to Hawaii we had no insurance. Our best insurance was careful navigation and thorough anchoring. From Hawaii to Alaska we managed to get a further low-cost rider. Normal coverage for the whole trip would have been some $2,000–$3,000 and way out of our budget.

Anchoring is a key safety issue. One can be a bit lax for short lunch stops. However, for overnight stops or when leaving the boat to go ashore for any period it should always be anchored as if a storm was on its way. We did have a couple of interesting problems in coral areas described later.

I think we were reasonably well prepared and ready to go. Any further preparation might have developed into some form of procrastination about leaving. It could also have been expensive: another cleat here, some more rope, spare batteries, more food, and on it goes with the bootline fading out of sight as fast as our bank balance. So we had set July 16 as the day, with no backing off even if we had to hide up around the corner in order to get some sleep. All those farewell parties were taking their toll.

On the subject of the bootline, on the last haul-out we had raised it 4". It had fast disappeared, not surprisingly I suppose, when the load is added – 125 gallons water (1,250 lbs), 85 gallons fuel (850 lbs), anchors and chain (500 lbs), food (300 lbs), books and personal gear (500 lbs), life-raft and dinghy (200 lbs), miscellaneous items (say 500 lbs), for a total of nearly 4,000 lbs. Two nights before we left, we had hauled her again and raised the bootline a further 2". It was not too straight a line, it being late at night, and painted in a hurried manner due to friends coming to see us.

I was beginning to get that light-headed feeling! The roaring noise down below was on the increase, the aft cabin was packed with kids, when lo and behold in motored *North Winds Five* from across the Burrard Inlet with the Sidney-smith family. They had just returned from a 2-year offshore trip and were itching to go again. Things were getting hectic. I explained to a friend where I had left our car, apologized for not getting a new tire (I had tried that morning), gave him the keys and wished him luck using it over the next year. I received with pleasure a drink thrust into my hand, posed for a photo (this could go to one's head), explained the route once again, checked my watch (why, with a spare year ahead?) and remembered that just a short three years ago and only a berth away we had had just such a crowd at the launching of *Sky One Hundred*.

Friends were still coming and going but it was time to get going. The rain was still coming down. I started up the motor and rolled up the rain cover. OK this was it; lots of hugs and kisses (should be doing more of this), tears here and there. "*Take care, take care.*" Final photo poses as we backed out of the dock, rudder hard over (better not muff this one), into forward gear and down the marina we went, past *North Winds Five* getting ready to follow us, a final picture back, a last wave back, and then facing forwards moving past the log booms where we have seen as many as 30 seals basking and smelling in the sun. Now we were aiming for Lions Gate Bridge with *North Winds Five* catching and passing us and leading the way out.

It was still raining, and we felt sympathy for Jeremy standing alone in the bow looking at his girlfriend, Linda, aboard *North Winds Five*. The smell of sulphurated hydrogen wafted over us from the sulphur piles of a loading terminal and then it was heads-up as we passed under the Lions Gate Bridge. The sweep of English Bay opened up, and what was *North Winds Five* doing? The yellow-slickered male members of the crew had moved onto their foredeck and had something behind their backs. Oh, beautiful! Conch shells from their trip; their mournful wail repeated itself in farewell blasts.

CHAPTER 4

Off to San Francisco

"I heard the sherry gurgling down"

Now, as *North Winds Five* turned back, we headed across the bay, past the Point Grey bell buoy, aiming across Georgia Strait to pass through Porlier Pass. Retreat Cove was our destination for the night. We beat our way across in the rain, Goldfinger worked fine and we were able to stay under our dodger. At the Cove we were happy to tie up, clean up the boat and relax. Good friends building a cottage there came aboard for a farewell drink and christened our guest book *"Westward Ho! (Ho! Ho!)"*

Next day, July 17th, was Jeremy's 16th birthday and Erica baked him a delicious chocolate cake. Heather and I were pleased that Erica had done this for her brother. In our new and close quarters for the next year we hoped their normal agreeable relationship would be maintained. However, one would never know; it was not like being in our house when they could roam far and wide or visit their friends. In fact, relationships between Heather and me and between all of us would have to be carefully preserved.

We cleared Canadian customs at Bedwell Harbour who listed our gear on an odd green card (2"x3") which the officer signed. I hoped that we would not lose it on our trip as it would be proof that our gear was bought in Canada. Another session with American customs at Friday Harbour, this time to obtain a six months cruising permit in the USA and we were free of red tape for a while. The customs here did get a little irate with Heather doing her shopping while we were at their dock! Off down to Anacortes where we stayed for a week getting organized.

Friends, Ozzi and Sandi from Mercer Island, brought up all the equipment we had bought earlier in the USA and this we proceeded to install. The main item was our ham radio. Its installation scared me somewhat due to all the dire warnings in the manual about reversed connections, serious damage with improper tuning of the aerial, possible severe electrical shocks, and so on. I had visions of ending up with a puff of smoke and a little pile of molten metal, that is, if I had not electrocuted myself first. We got an "expert" in to check my installation which was OK except for some fine tuning related to the aerial length. He then gave us a quick lesson on how to use the radio while Heather took shorthand notes which we could refer to later.

Readers might be questioning our often late and minimal gathering of data for our radio operation, celestial navigation and many other issues – were we negligent? Were we competent to undertake this trip? Good thoughts, but there was a need to keep moving on for, if we attended more diligently and earlier to all the things we should do and should know, we might not have started the trip for another year. Some boat builders we knew were still building after we returned. One builder, who had leased land for a year on which to build his boat, took six months putting up a shed, and, had just started building his dream when the owner wanted his land back. We expected to learn about many issues as we went along. Setting a time target was an excellent way to highlight progress towards one's intended completion; without it, time and progress could just drift along.

Following a 30 nautical mile sail to Port Townsend we had a splendid reach, rail down, in calm water in front of this town which dated back to at least 1860. In Port Angeles a rather uncooperative marina staff left us juggling our boat around in a good breeze so, at the invitation of the owners Bo and Joyce, we tied up alongside their 66' power boat, the *Twin Isles*. We invited them over for a drink and then we viewed their boat. We did a big shop and on leaving the next day just got a "click click click" on starting the engine. The batteries were a bit low but not that low. I removed the starter, took it apart, cleaned all connections and lubricated all parts before reinstalling it. Eureka, it worked.

The good breeze was still blowing so we went to play frisbee on the beach. A bad throw and the wind took it out to sea where it quickly drifted away. Jeremy and I rushed back to our boat, grabbed our dinghy and rowed after it. We retrieved the frisbee about a quarter of a mile offshore. The wind was blowing across the bay and while I had to row furiously for 10 or 15 minutes to get back, I thought once or twice we would end up on the far side of the bay. I was pleased I had bought big standard wooden oars that fitted into the large rowlocks on our Avon dinghy. With any other dinghy, fitted with little plastic oars and delicate fittings, we would have never made it. The wooden oars could always be replaced whereas broken plastic fittings would be a problem.

All of our sailing had been done on the relatively calm

waters behind Vancouver Island amongst the local islands. While Erica had said earlier, *"I always loved being on a boat whether it was Sky One Hundred or Kolus before her. I enjoyed the independence when we arrived somewhere and exploring the different islands and beaches"*, we were now beginning to experience slightly rougher seas and sea sickness. Conditions changed when we left for Callum Bay and found ourselves motoring all day against a strong headwind and nasty irregular seas. Erica lambasted us continually that she did not like this at all, she was going to be sick, it was alright for us but she would rather be on land with horses, she had never wanted this and so on. It was not fun, the weather or the tirade, but we just kept going while still feeling guilty about our proposed trip.

Erica said, *"In retrospect, it was probably a good thing as nobody was to find out just how seasick I would get on the ocean swell! By the time they did, it was too late as we were now off the NW coast of the USA on our way to San Francisco with no turning back! A great worry for Mum and Dad and probably Jeremy too but for me, it was a horror that didn't stop for 7 days. Even as we were flying under the Golden Gate Bridge, I was still cuddling my bucket, taking it out on everyone around me as to how it was unfair . . . Our stay in the Bay of San Francisco and seeing all the new sights, exploring new places with not a wave in sight was great for pushing all thoughts of sea sickness to the back of my mind"*.

However, we were impressed by a U.S. navy hydrofoil which passed us high out of the water at about forty knots – the only way to travel. At Callum Bay we declined paying $10 for a berth and anchored out in the bay. Next day we headed for Neah Bay, our jumping off spot into the North Pacific waters.

Our sextant sun shots over the past few days had been providing some spectacular ranges of location, the worst being somewhere in Mexico. However, we were getting better at determining our position and were generally accurate to within four miles, which was a relief. We were impressed with the weather service on our VHF radio.

At Neah Bay, a charming customs lady came aboard, had a cup of tea and told us our U.S. cruising permit had been improperly prepared, made out a new one and advised that it was not mandatory to check in at all ports of call after San Francisco. We tied up on some log booms and waited for some favourable winds. One thing we learnt about official institutions was that rules for the same requirement often changed from official to official.

We kept working on our sun shots but after three days of waiting for a decent wind I started to think I was getting cold feet about departing into the Pacific Ocean and all that we might experience. Finally the weather forecast indicated the south west wind would change to north west. So on Friday, July 29, we got going. I don't believe traditionalists ever leave on a Friday as, from myths, it is considered the unluckiest day of the week. And leaving on a Friday the 13th, 13 being unlucky too, would likely induce some disaster or shipwreck. A study of ships sunk in the Great Lakes has shown that the greatest number sunk left port on a Friday. However, Heather and I were married on a 13th day and we're still sailing! As we moved out of the harbour we called up U.S. customs and advised them of our departure and they wished us good luck. We gave the menacing Tatoosh Isle and Duncan Rock, off Cape Flattery, a good clearance and, somewhat close-hauled, headed on 210-degree course. It was not the best of sailing and we stayed this way for two days until the wind eased to the north and gave us some respite from the twisting and jolting.

Land soon disappeared from sight and we were alone on the ocean, a new experience that we were going to have to get used to. We had heard that some people have not liked this experience because they feel there is a loss of control but perhaps it is also the feeling of loneliness or even some form of agoraphobia. Heather's comment was that we wouldn't run into anything! At the time, I did not have any particular feelings, my thoughts were mainly on sailing the boat, the next sun shot, what the weather might be and, of course, how Erica might adjust. A school of at least a hundred porpoises moved past at high speed and we thought of 'Jaws' as a shark whipped around in the water for a while.

Unfortunately, five hours into our first day – due to the somewhat violent movement – our 135% genoa split a seam with an unpleasant bang. This was not a good start for twelve months of sailing. But we still made good progress at 5-7 knots with our lapper sail. I wished I had requested this be made 1-2 ozs. heavier. During the day, we saw dolphins, sharks, and a couple of freighters, and, at twilight, a ketch with brown sails far away on the horizon to the east. It was, as we discovered later, a solo sailor from our club, Colin Hempsall, aboard *Moon Island* and on his way to Hawaii. Although it was sunny to cloudy we wore sweaters, boots, floater jackets and oilskins.

Around midday we managed to get a quick angle on the sun before it disappeared for the day. Before the trip, all position calculations had been carried out sitting at nice comfortable and stable tables with absolutely no movement. Now I had my first try at working out our position while underway. What a ghastly experience, sitting at our table with my legs braced out trying to hold myself in place while books slid about and figures from various navigation tables kept leaping up towards my face with the twisting boat motions . Fortunately, I managed to hang on and complete a line of position which generally agreed with our dead-reckoned position before I rushed on deck to discharge my day's calorific intake to leeward. I felt better after that and was not sick again until the start of our next trip.

Erica, however, was not feeling good at all and was crying and saying we should turn back. Heather was getting quite concerned about Erica's continued state of discomfort especially as she did not eat anything. She had cried on several occasions before we got to Neah Bay and I was beginning to feel a real heel taking her off like this for my own benefit. Between her sniffles I was again told, *" It's alright for you, you like sailing. . . "*, which did not ease my conscience one bit. As the day progressed we all found sailing along rather boring and, surprisingly, were asking each other if this was really what we wanted to do for the next year. It must have been quite a low moment for all of us as we even got as far as discussing selling the boat and spending a year in Mexico. I wondered how we could get to even consider such a switch to our plans which we had been working on for so long. We later asked other yachties whether they had had similar doubts and quite a few said they'd had remarkably similar reactions and discussions.

At the end of the first day we donned our life harnesses, which were clipped on inside the cabin so that no one exited the cabin unclipped. Whether it was nerves or the feeling of security that being together produced, or whether the main cabin area was the most comfortable part of the boat, that was where we spent the next few days of the trip and to heck with separate bunks.

Our sleeping quarters were not what we had anticipated but were a matter of convenience as they were in the most stable locations. Our long couch built into the shape of the hull was very cosy, and on it slept Erica and a person not on watch. The other person not on watch slept on a mattress on the cabin floor alongside, all a bit of a pig-sty arrangement but comfortable, and Erica provided excellent body warmth. Heather wrote in her Newsletter home, *"Patrick was our en-*

tertainment when coming off watch as he just dived below, unclipped his harness and, still wearing oil skins, boots and hat just slumped into the waiting warm sleeping bag. Within seconds snores could be heard from above. How envious I was of his ability to relax so easily but it was probably more like exhaustion thinking about the family and boat, not to mention determining our position".

Watch followed watch without incident, and the watch person could sit out of the weather under the dodger but for an occasional check of the compass and for ships. Our fifth crew member, Goldfinger, was doing an excellent job.

Our first dawn saw us beating through 4 – 5 foot waves but fortunately not the short choppy variety of Georgia Strait. How Heather produced a breakfast of porridge, bacon, and eggs I will never know. More interestingly was how we bolted them down – did we think this was going to be our last meal, was this a nervous reaction or were we just plain hungry? Although it was sunny with 5/10's cloud (50% cloud cover) the weather was cool and we still wore sweaters, boots, floater jackets, and oilskins; it was very difficult to get warm again after getting cold as we had no heater. We had a visit from a small bird which just sat on the bread Jeremy fed him. Out of nowhere large albatross-like birds appeared swooping along the crest of the waves and disappearing in the troughs.

The second night was clear and bright. I took some star shots which provided a specific position by triangulation. By the third morning the wind was swinging to northwest and I reckoned we were some 160 miles offshore and hopefully clear of any shipping lanes near the Oregon coast. We saw no ships after leaving Neah Bay until we reached San Francisco.

When in sight of land, our focus was usually on the land and the horizon. Being completely out of sight of land though after some days I found it somewhat disconcerting:

> "We look 360 degrees around the horizon and there is nothing. The predominant colour is grey; grey clouds and grey sea. The greys of the moving water are laced with the flash of whitecaps. There seems to be a complete randomness of wave movement. Sometimes there is sudden thump on the side of the hull which makes me jump or there is a rush of water seething and hissing past the stern. What the heck are we doing out here? The enormity of our plan strikes deeply into my mind and body for the first time. Perhaps Erica is right. I look again around the horizon but there is nothing. I console myself with the knowledge that this boat is strong and we have a radio and a life raft".

Later, as we became more at ease in this new environment, we tended to focus more within the boat.

As we sailed southwards down the coast in relative comfort I could not help thinking that in 1778, Captain Cook sailed into Nootka Sound, two hundred years earlier and Sir Francis Drake, the scourge of the Spanish, sailed north as far as Washington State and maybe further in 1579; adventurers indeed. While working in Colombia, I was told that when the Spanish were ransacking the hinterland for gold they would load their ships in Cartagena and then sail forth to Spain. Unfortunately Drake and his merry men would be waiting to board the ships and take the gold. To prevent more of this, the Spanish built the Canal del Dique from the Magdelana River to allow them to sail from a point further south on the coast and hopefully avoid Drake's ships. It was probably a Brit who told me!

> "We were shooting down the wave fronts and whooping with exhilaration..."

The continual motion of the boat was being accepted by my sensitivities but not to the extent that I was comfortable working down below. In fact none of us were comfortable below. All we wanted to do was eat, sleep and sail. Helming was no great chore as it helped to pass the time. None of us wanted to read, and navigation calculations and taking sun and star shots were tasks I could well do without. Paddington Bear, our yellow-hatted mascot attached to the binnacle, maintained his smile.

At this time I was having difficulty reconciling the differences between our position as determined by my dead reckoning versus our calculated celestial position. Climbing onto the coach roof to see the horizon better while the boat is rising, rocking and falling, it was difficult and frustrating to adjust the sextant to hit the edge of the sun and the horizon at the same time. Global positioning instruments were not on the market yet.

The horizon never seemed to be smooth and there was a lot of movement on *Sky One Hundred*. This was nothing like standing on a static beach in Vancouver and measuring the angle to the sun. While I felt confident of the measurements, was I making serious errors because the celestial position showed us some 20-30 miles further offshore? The noon shots were very frustrating as, after measuring the sun reaching its zenith and then descending, the sun would start rising again, and then, after a while, start down again. What was going on here? After 40 minutes of readings my arms were aching, my stomach nauseous and my temper foul. Heather, who was taking the time for each reading, was feeling much the same as myself. This same "up and down" happened on the next day and I suspected the black plastic frame of the sextant was perhaps warping in the sun. The two sextants had 8 and 12 minute errors, and I was later to appreciate one had a lateral error. I bought a proper metal sextant in San Diego and had no further problems.

The wind was getting stronger and on the third day Goldfinger would no longer steer the boat reliably and we disconnected it. The wind had swung to the northwest and had increased to 15-20 knots. While the waves were increasing in height steering the boat manually was fun. We were shooting down the wave fronts and whooping with exhilaration when we hit speeds not reached before. *Sky One Hundred* would slow and pause as we lost the full strength of the wind in the troughs before the next wave lifted us up again into the full wind and we accelerated and took off again down the next wave front. Who said sailing wasn't fun? With Heather, Jeremy, and I helming in turn, which we definitely wanted to do, we would each try to exceed our top speed of nearly 15 knots. While the seas seemed quite mountainous we rode them with ease. It was pretty exciting!

At this time Erica was still quite sick and was not eating and not wanting to drink. Heather was very worried about her and as a result was not sleeping. As Heather came off watch for the night, I passed her a bottle of Bristol Cream sherry and advised her to have a good slug to calm herself. She put the bottle to her lips and I heard the sherry gurgling down. It worked.

I later realized why Goldfinger was not working. I had set up the main rudder to balance the boat but if the wind speed changed its bigger area would overpower that of the auxiliary rudder operated by Goldfinger. In other words, I was not balancing the boat correctly. I had no further problems once I realized that the boat must be balanced so that the main rudder was always in line in a central position. Double-reefing the mainsail also helped.

That night with the wind increasing to 20-25 knots we triple-reefed the main. Before nightfall, I hanked on the storm jib and connected the sheets just in case it blew up some more. Well, it did, so I changed the lapper for the storm jib. By morning it was blowing 25-30 knots with the seas about

12-15 feet. It was difficult to estimate wave heights but this was ours and it may have been low. Many times on our trip we heard sailors saying they had been in 20-30 foot waves.

"this one WILL drop in the cockpit"

The effort to change the forward sail could be an amble forward on a nice day, or it could be a bit of a fight on a heaving deck at night in strong winds. With the lifeline clipped to the safety line on the deck, one hung onto the handrail on the coach roof to get to the mast; at that point the helmsperson would bring the boat into the wind so that when the wire halyard quick release was let go the foresail would drop onto the deck. At this point it was necessary to sometimes crawl to the bow, un-hank the sail, unhook the halyard, untie the sheet, hank on the smaller sail, hook or tie on the halyard, re-tie the sheet onto the smaller sail, tie down the loose sail on the deck, return to the mast and haul up the sail when the course could be resumed. Quite simple usually, but often done by touch in the dark or near dark. Sometimes I wished we had an inner foresail or cutter rig so that the inner foresail could be hauled up quickly and the big foresail dropped. It would have made life easier.

The wind increased again in the morning so down came the main and we spent all day using only the storm jib, racing down the wave fronts surfing at 10 and 11 knots. At times we were hitting close to 15 knots again. It was absolutely exhilarating but tiring. *Sky One Hundred* performed well and gave us a lot of confidence as she lifted her stern up over each wave. The waves seemed to be getting bigger and looked as if they would drop into the cockpit. Sometimes as we looked up at a following wave which seemed well above us we'd call out *"this one WILL drop in the cockpit"*. They never did and we gained more confidence as each wave slid under our stern.

Heather said, *"She really felt as if she was riding a beautiful race horse in a steeplechase while feeling like the Queen of the Pacific"*. I felt I was a lucky man to have such a wife and companion who could adapt so positively to these offshore conditions. Often as we rose up on the peak of a wave we would get a quick look around at the open sea. It was quite concerning to see the roughness of the surface. Sometimes there would be a big black block of water standing up above the horizon leaving me wondering if it was a boat of some sort. Fortunately, we did not encounter any 'rogue' waves that were extra big and could come from an unwanted and surprise direction.

By evening we reckoned it was blowing 40 knots. This speed (80 kms/hr) was just touching *"strong gale force"*. Putting one's head outside of a car going that speed would be quite discomforting; it was time to ease off for the night. I asked Heather to check in Adlard Cole's "Heavy Weather Sailing" about heaving-to. After pulling our jib across to the windward side with the mainsail doused, the noise and motion eased off as we edged along at 1–2 knots across the waves, occasionally heading up towards the wind and the wave crest and then dropping away again. It was surprisingly peaceful with the reduced speed. Once we had got used to the odd thump or two of waves on the hull with water falling on the deck we managed to catnap till morning. A look into our cabin would have shown a pile of bodies lying in the main cabin. When thinking about going offshore before this trip, I used to read the "heavy weather sailing" book in bed at night. There were horrific pictures in it of huge seas which Heather had said she did not wish to see as they would give her nightmares!

On the fifth day it was still blowing but we put up the main, triple-reefed, and were off again on another switchback ride. Heather still managed her cooking skills in all this, coming up with soups, fried eggs, sandwiches and hot meals which kept us happy and going strong. Erica was nibbling at dry biscuits and sipping water but not keeping it down, which continued to worry us. Chewing gum seemed to relieve nausea. Jeremy resorted to the occasional Dramamine sea-sickness pill but they tended to make him drowsy.

Goldfinger was performing OK but the brackets had started a slight movement as the bolts worked the holes in the fibreglass bigger due to the big forces from the auxiliary rudder. This required a miserable interlude lying flat up in the transom bunk tightening the bolts. Later, I was advised the toilet was bunged up – too much paper – and so I had another pleasant interlude taking it apart and cleaning up the mess. While it was overcast we did manage some sun shots and during the afternoon the wind dropped to 10–15 knots.

During the sixth day, the wind died and the sea calmed down. Erica came on deck for the first time but she was pale and thin and her jeans were held up by a rope. Sharks were sighted, and I apparently provided a source of hysterical laughter as I rolled to and fro on the cockpit floor asleep like a corpse. It was too overcast for any sun shots. Towards the day's end the wind dropped completely and we motored; in black nothingness later that night, this was quite spooky. My imagination played tricks in believing that we would run smack into an unseen object or, worse still, drop off the edge of the sea. Without an auto pilot my course would wander significantly with inattention to the compass. There was no starlight, no shadows and nothing visible to aim at; however, after a while, I became very conscious of the motion of the waves and their direction which helped to keep me on course. On stopping the motor once we just rolled slightly in the swell. There was no sound except for some birds — what a contrast from the earlier days. We put the tri-light on at the top of the mast and birds were attracted, some hitting the sails and sliding down onto the deck. What were they doing so far off shore?

On the last two days we had only seen the sun once, and briefly at that. With no stars either I was unable to confirm our position. Our radio direction finder was not picking up any radio beacons although San Francisco radio stations were strong and gave us a general direction. We started heading into the coast at a 120-degree bearing. We saw one light passing in the distance. Soon we were sailing again and did so all throughout the following day in about 15 knots of wind. It was very dull and overcast and sun and star shots were not possible. By now I had a distinct feeling that we should have picked up the Farralon radio beacon or light off San Francisco and so altered course to 90 degrees. At 2300 hours, Jeremy saw a glow in the sky for a short moment and believed it was San Francisco. An hour later we picked up two lights, one blinking at the same frequency as Farralon. There should have been three lights so we decided to heave to and wait for daylight. It was an uncomfortable night.

At dawn we motored towards the light only to find it was a strobe light on a fish boat. We called them up on our VHF radio only to be told we were 20 miles **south** of San Francisco! Damn! We had no doubt drifted south some 15 miles over night. Our latitude had looked alright but the incorrect longitude rather indicated I should not have paid so much attention to my dead reckoning in the first few days out. We had a super reach into shore and finally grey shadows of cliffs and hills loomed up and resolved into valleys, roads, and houses. With much relief the top of the Golden Gate Bridge could be seen. At 2200 we were surging towards the bridge at 10 knots and more with a triple-reefed main. In the darkness we came surfing in under the huge span of the bridge and were ecstatic in completing the first leg of our trip. We motored into a marina next to the St. Francis Yacht Club and grabbed the first berth we could find which happened to be the Harbour-master's Dock. We did high-fives all round and then I called up the Customs and called home to advise of our arrival. We dropped into our bunks for a peaceful sleep. We had been on the move for eight days.

Next morning, Saturday, San Francisco impacted itself upon us. With wind surfers flashing past *Sky One Hundred* and boats madly sailing in and out of the marina in a 20-knot breeze, along with joggers, kite flyers, visitors, and musicians it all presented a lively scene. Some boaters, as they passed by and seeing our Canadian flag, would ask if we had had a good trip down and congratulated us. This, and their obvious friendliness, was very refreshing. Erica perked up under all this attention (maybe this was when we first called her "Perky") and seemed more interested in continuing the trip.

We stayed for three weeks and enjoyed every moment, visiting Fisherman's Wharf, the Maritime Museum, Art Gallery and Science complex, Golden Gate Park, the zoo and, of course, riding the cable-cars; our feet and backs ached. Sailing in the Bay was superb where the winds were strong but the waters more or less calm. We explored the vast harbour, cruising close to Alcatraz, under the huge Oakland Bridge, to Tiburon, to Sausalito, and to Ayala Cove on Angel Island. This cove is the equivalent of Snug Cove on Bowen Island. For some exercise we climbed the 788' high Mt. Caroline Livermore peak for a 360-degree view of the harbour. We spent three free nights at the St. Francis Y.C. (15 cents/foot per night thereafter) and also spent a few days at the Golden Gate Y.C. We took buses to visit down town and hired a car for a few days to explore the harbour area, the Golden Gate Bridge and drove into the hills behind Oakland.

While in the San Francisco area, I also added a second boom vang or preventer. This was a rope leverage system by which the boom could be hauled down tight from the cockpit. Its function was to keep the boom from kicking upwards in strong winds which helped us go faster. When sailing with a single vang, I would, for better use, unhook it from the bottom of the mast and attach it to a mooring cleat at the side of the boat; this would prevent the boom from wildly and dangerously sweeping across the cockpit to the other tack if the wind managed to get behind the mainsail. When we changed tack it was necessary to go forward and re-cleat the vang on the other side. This was an unwanted task, so I bought rope and sheaves and made a second vang to have one on both sides. The lines were slightly stretchy and were attached to our spinnaker winches which were not being used. This double vang system had many advantages including allowing a controlled gybe in high winds.

We sailed up the Sacramento-San Joaquin River Delta on the Sacramento River, a fantastic area of inland waterways with a large dredged channel that allowed 50,000 tonne freighters to reach Stockton 80 miles inland. We barely touched the fringe of this waterway finding on a couple of occasions that we had a tendency to do our own dredging.

The first time was in the Pittsburg Marina where we had been given a berth. On leaving the next day Jeremy went to push the bow out so that we could motor away but the boat did not move. Heather jumped down and helped Jeremy push but still nothing happened. I looked at the depth sounder which showed 3' (our draft was 6') – a very new experience!! I could not motor out as we seemed to have set up in fine silt. To get the boat out we had to take a line across the exit channel to a dock opposite and proceed to winch our bow out sideways – a very slow process. Directly the bow was clear of the boat in front of us I started the motor and found we could just motor slowly, very slowly, through the silt into deep water. The irony of this was that earlier we had walked around to the marina's gas station to see if they were open and were advised to watch out for a certain low spot just in front of its dock!

The second occasion occurred in the freighter channel to Stockton when I decided to chintz on a bend and cut inside a marker assuming, very incorrectly, there would be some spare depth. We had been going downwind at about 5 knots when there was a disturbing "soft" slowing of the boat. Heather, who had been standing at the companionway looking aft, disappeared, followed by a crash as she made intimate contact with the forward bulkhead. I immediately and simultaneously turned towards the channel, while ripping the boom across into a reaching position and slamming the motor on at full revs, knowing, that if we stopped, the silt would likely set up around us and we would be stuck. We managed to move back into the channel at about 1 knot or less. Phew! A narrow squeak! I was then the target, the very direct target, of quite a bit of squawking about my lack of sailing expertise, from a bruised and irate Heather.

"Beware of seagulls standing on water" became our alert for spotting shallow waters and sandbanks in the area. However, we decided to ease back to deeper waters and stopped at the Vallejo Y.C. after passing some 137 warships, all mothballed. This was a really friendly club where we were lent a car to do groceries, visit the vineyards of the Napa Valley, sample some excellent wines, and see the countryside. After the second vineyard visit, we managed to avoid, at further vineyard visits, the usual introductory talk and slide straight into the next testing group! On a return from one trip we saw a club member was having a lot of trouble fibre-glassing his cabin top. It was in such a mess with resin dripping all over his teak work that I helped clean it up and finish his job.

Just before we left we witnessed what turned out to be somewhat of a disaster. A boat, under full sail, came into the club's rather small harbour. There was a man and a boy on board. The boat came in downwind and then turned through

180 degrees, slowing as it moved towards an empty berth. We thought this was going to be a neat bit of sailing but, surprisingly, it continued to turn through a further 180 degrees picking up its earlier course. We wondered what the man was going to do now as the boat gathered speed going straight for the clubhouse. He ran forward on the deck (for what reason?), tripped, and fell flat, just as the boat's bow hit a dock directly in front of the clubhouse, rose up and stopped with the sails flapping. We wondered what the members thought who were watching from the clubhouse windows. It was an extraordinary demonstration!

Jeremy and Erica, who had done virtually no school work on the trip, had, since our arrival, been putting in some study hours but it was tough on them – there were so many distractions. To undertake a correspondence course under normal circumstances takes a lot of effort and a good measure of self-discipline. So to see them settling into their tasks, usually without our urging, was always pleasing to Heather and myself and we would help them however we could. Perhaps part of their motivation was to keep on a par with their friends back in Vancouver! Strangely, during our time in San Fransisco, very little reference was made to our earlier thoughts at sea, to possibly change our ocean voyage plans and spend a year in Mexico. I thought the wonderful time we all experienced in the recent days was possibly viewed as an indicator of what the rest of the trip might be like.

Finally, back to San Francisco where we met Vancouver boats *Snark 11*, a junk rigged ketch sailed by the Breckner family, and a CT41 named *Moon Island* with single-hander Colin Hempsall from our yacht club aboard. It was his boat we had seen on the first day out. He had had to divert from his course to Hawaii as his wind vane had broken. Over a few drinks he also told us he had second thoughts about his sailing trip after just a few days out. We also met Rex McDowell, aboard *Rascal*; he urged us to go to Alaska on the way home. After 18 days in this exciting harbour and city it was time to be moving on, and, with another load of food, we cleared the Golden Gate Bridge and turned south for our coastal hopping to Mexico.

CHAPTER 5

Harbour Hopping to San Diego

"This action would signal a missile strike that would be the pride of any military group."

We departed San Francisco on August 26 to start coastal hopping down to San Diego. We did not anticipate leaving there for Mexico until the end of October. By this time we understood that most cyclonic activity, such as violent squalls (*chubascos*) with thunder and lightning, encountered during the rainy season to the south, should have ceased. Allowing about three weeks for final preparation in San Diego (the last comprehensive supply source before Hawaii), we felt 6–7 weeks to cover some 500 miles would provide some nice and easy cruising.

We hugged the coast through to Santa Barbara stopping

at Half Moon Bay, Santa Cruz, Monterey, Stillwater Cove, Simeon Bay, Morro Bay and passing Point Conception. Southwards to San Diego we stopped at Ventura, Port Hueneme, Santa Monica, Los Angeles, Redondo Beach, Newport Beach, Dana Point, Oceanside and Del Mar.

Coastal harbour hopping with its many stops was a relaxed and delightful way of cruising. We never knew what adventures might happen next. Except for one night, we always made the next harbour before nightfall. Supplies were always available. There were many different sights to see as well as people and boats to meet. On many occasions when arriving we found locals coming up to talk to us, to take us around, lend us their car, offer to play tennis and even take us out to dinner. It was an incredible lesson in friendliness and hospitality. Jeremy and Erica responded well to these overtures. This was a relief to Heather and myself because there were always adults for us to mix with but not so many children and ours were always being uprooted to go onto the next stop just when they might have made some friends.

Fortunately we were always criss-crossing with boats on similar ventures so that we were able to make loose arrangements to meet up again. The two boats we mainly travelled with were *Active Light* from Port Townsend with Penny MacInnes and Dick McCurdy and their friend, Dave (who we met in San Francisco), and *Restless Wind* from Seattle with Gerry and Randi Jacobs and their three children: JJ, Renee, and Rachelle.

Travelling more or less together provided the opportunity for entertaining and social comfort, practical support as well as potential collective security although the latter was hardly required. Dick had built his boat and Penny, was an English girl he had known; Dave travelled with them from time to time. They were all good fun. They actually parted company from us in San Diego sailing direct to the Marquesas. We met up with them in Papeete and then they carried on around the world. Gerry and Randi were a super couple travelling in their Morgan 42. We finally split with them in Kauai where they returned to Seattle while we travelled on up to Alaska. We were pleased we blended so well, particularly their children with ours. At any time the children could leave and go off together and leave the parents.

During our last two days we stayed at the St Francis Y.C. There were thick swirling mists and it was not a time to leave. The children were doing their school work and at times we listened in on the ham radio to Colin Hempsall talking to his wife as he headed for Hawaii. Heather and I visited Fisherman's Wharf where we purchased a 4 HP outboard motor for our dinghy. The next day we filled up with diesel fuel and left for a short day trip past St. Pedro Point and anchored in Half Moon Bay for a quiet evening. Well, it was not all that quiet as Jeremy tried out the new motor, flying around at full speed.

After another short sail to Santa Cruz we tied up with *Restless Wind* at the yacht club. We all spent a great day on the superb beach and enjoyed the pier and boardwalk. At the fantastic fairground we dispersed any random "aggros" on the "dodge 'ems", a misused description for bumper cars if ever there was one, as my idea was to charge any car in my way, as I had done as a kid in England. Heather and Erica enjoyed the Haunted Train ride. In the evening we went up to the club's party and dance that was being held to celebrate *Merlin's* win in the trans-Pacific race, while Jeremy tried to call his girlfriend.

Coming into Monterey from Santa Cruz across a wide coastal indentation, we sailed into a sea mist for much of the way which was very disconcerting. Navigation was by dead reckoning, a radio beam and soundings. While a challenge to reach your destination, sailing in a dense white fog or mist was not a scene that one relished; travelling with the motor on was most unacceptable being unable to hear any thing or any approaching boat.

Still it was better than being in the Great London Smog of December 1952. This five day smog was due to excessive winter burning of house coal fires. The smog was so dense that often one could not see one's feet. Even an indoor show was cancelled because the audience could not see the stage properly. The particulate in the yellow smog made one's face black where not covered by a scarf. White shirts became grey very quickly. Thousands of people died from lung problems. Transportation was slowed as bus drivers had to be led by their conductor holding up a flare and leading the way, walking on the curb of a sidewalk. Having recently moved, Heather's house had no phone yet. If she wanted to call me she had to leave her house and walk a hundred yards feeling her way to a public phone box. She was guided by her hand touching the front fences of the houses. Such dedication and love – I was indeed a fortunate fellow! Jeremy who later worked in the filthy atmosphere of a Chinese shipyard building a large cruising sailboat, once told his translator that the boat's destination was the South Seas and showed her a picture of blue sky and seas. Her response was "*No, that is propaganda, all sky grey.*"

Nearing Monterey I was concerned we were close to shore, especially as the depth was dropping rapidly and we still could not see the marina or the town. Just when we were about to drop the sails and motor (something I do not like as I would not be able to hear another boat) the marina loomed out of the mist. Luckily we got an inner berth and later found Harold and Ursula Schnetzler there in their L36 Lapworth, *Maradea*. They were from our yacht club and were doing a similar year's trip! We dined out, watched the crowds, sampled free wines and explored Cannery Row where old warehouse after warehouse had been converted into boutiques and souvenir shops – a back-aching and foot-killing process.

Stillwater Cove was a gorgeous spot just north of Carmel. Although not particularly still, it delighted us with its white beach, its hundreds of seals, some popping up to inspect us, and the multitudes of pelicans diving down for fish with a great plop into the sea, emerging with mouths bulging like over-filled plastic shopping bags. Really cute were the sea otters that floated by on their backs cracking shells on their tummies with a rock. We were lying close to a large kelp bed opposite the 17th hole of the Pebble Beach golf course. Often we could see golfers looking over the cliff for their errant balls; Jeremy and Erica searched for many on the beach. We walked around the course, admired the luxurious homes, watched a doubles match at the tennis club and wished we had our togs with us.

In the evening, Jeremy and I played frisbee barefooted on the immaculate green fairway. The following day we walked into Carmel and ogled with the tourists at a film being made in the main street. On return we found *Restless Wind* moored next to us. They came aboard and we all got more acquainted with this very easy going family. We delayed our departure and stayed for a BBQ with them on the beach, the first of many we would have together.

We left early the next morning for San Simeon Bay and after a rapid sail dropped anchor in "Hearstland". Jeremy, Erica, and I immediately pumped up the dinghy and rowed to shore. Heather, in an inspirational and visionary moment, put the kettle on and waited for our rapid return. Our first surf ashore was not successful. While quite innocuous looking, the surf played havoc with our mistimed run onto the beach and we were dumped into the cold water with Erica screaming blue murder. Heather had seen it all happen from the boat and had hot drinks ready. We learned quickly we must row in behind the wave and not in front. Another boat there had clothes out to dry – perhaps they had the same experience. We were agog with Hearst Castle, its architecture, blue mosaic pool, art and statues. The swimming pool with its royal blue and gold tiles would be a great place for a party! An amazing extravaganza but rather like a museum. Back in the cove *Restless Wind* and *Active Light* had arrived, so there was another BBQ on the beach.

During the quieter moments of our trip Jeremy and Erica were pretty diligent attacking the B.C. Correspondence course – at least Jeremy says he was! They wanted to keep ahead so that they could coast a little when we arrived at more exciting places. Dave off *Active Light* was a great help to Jeremy with the "new" math, although Heather and I never saw any problem with the old math. It was our view that core values were knowing "times tables" and the ability to play cards, especially "crib". It always beats me to see a bank teller use an adding machine to sum up two or three numbers or do a simple multiplication.

After our visit to Hearst Castle we waved goodbye to our boating friends and with a good breeze headed south for Morro Bay. At 7.15pm we just made it past the Gibraltar-like rock and into the harbour before the fog socked in. A boat thirty minutes behind us had to come in on a compass bearing. Care was needed at the entrance due to the continued shifting of the bottom contours with silting. We moored at the pleasant and friendly yacht club and avoided being on a buoy swinging in the tidal current. We tried to advise Customs of our arrival but could not make any contact. Next day a young guy was looking at our boat and we invited him, Patrick, aboard. The following day Patrick and Linda came over for a chat, organized tennis and shopping for us, took us to their house and gave Jeremy an old surf board. Later we played tennis with them and had a BBQ on the beach.

Meeting locals on this casual basis was such a splendid low-key way of enlarging one's understanding of other people, how they lived, their views on the world scene and politics, apart from being introduced to local scenes that a tourist might not normally see; a special beach, walk or restaurant. Jeremy and Erica were always ready to hear views other than ours, for a change, and would not hesitate to question differences of opinion.

This was a delightful place. Boats moored in line in the channel would turn about in the changing tides like troops to the order of the currents. Mist and fog were frequent and provided interesting and varying scenes distorting the normal perspective of our moorage area. The bay had a natural protection of sand dunes adding to the laid back feeling and interest to the harbour. The waterfront was a fine place with loads of fish restaurants and souvenir shops; it looked like a developer's paradise.. It was sad to leave our new friends and perhaps *Sky One Hundred* felt that too since, as we tried to leave the dock for Santa Barbara, I found there was no propeller action. Dammit! The coupling to the engine had come loose and I had to retighten it. This was to occur later in our trip with the need for rapid response action.

Point Conception was said to be a windy spot. One racing boat we met said they had experienced 45 knot winds on the way north. Fortunately we sailed past in moderate winds. Our friends, Patrick and Linda, had said during supper on our boat, that *"Past Point Conception there will be sun"*; they were right and we experienced sun right down to San Diego. Off came sweaters and floater jackets and Heather would type newsletters home while in the cockpit in her bikini. Our long time friends, George and Ursula, had given us a block of paper and each sheet was headed in one corner with a map showing where we were going, and on the other corner, a flag with "Hepajeric" written on. As we travelled we could write

"Newsletters" and show where we were on the map – a neat idea which we much appreciated. Heather was the writer of most of the Newsletters; I would sometimes add "Captain's Comments" and much of her writings are embodied in this trip story.

After Point Conception we encountered some oil slicks among the offshore oil rigs but understood they were natural leakages. We saw the occasional shark which Jeremy would try to bombard with old tennis balls. We did not make Santa Barbara due to heavy winds and some fog and elected to pull in around Point Sal and dropped anchor in 7 fathoms behind a small rock island at about 6pm. We were up at 11pm and 3am considering leaving as upwind of us was our main problem. It was a rock covered with seals or walruses that spent the night honking for terrestrial rights or perhaps mates – we didn't know and didn't care because their smell was atrocious. We left at 5.30am.

At Santa Barbara we followed the channel entrance markers and just made it in before total darkness fell, tied up alongside *Restless Wind* and had a very necessary reunion drink. The next day we investigated the yacht club. We found one of the girls off *Whistler Wing IV* looking for a stiff drink having just run a fashion show in which the male model had calmly walked on wearing only a blazer.

The fascinating town boasted beautiful Spanish architecture along with an interplay of palm trees, colourful mosaic tiles, and cool courtyards which was very pleasant and refreshing. We went to an architecturally interesting movie house and saw the "Starwars" premiere. The inside of the attractive theatre was designed as if one were sitting in a Spanish square surrounded by pretty balconied houses with a backdrop of snow-covered mountains and stars in the sky.

We played a few games of tennis on excellent flood-lit courts with their very efficient booking system; Erica was improving and was beginning to look quite leggy. There were many racing boats here and Heather and I would wonder if Jeremy might jump ship for some wilder experiences. Dick and Penny on *Active Light* arrived and, much to our surprise, we found they had bought a parrot, complete with cage and leather glove (ouch! —steel would be better!). They called it Mybird and hoped to train it to speak. We all believed it had a deficiency in that direction because six months and four gloves later it could only squawk and litter – the latter with considerable variety.

We sailed the 25 miles across to Santa Cruz Island along with *Restless Wind* and *Active Light*, passing giant oil rigs on the way. The Channel Islands were quite barren with lava-like terrain covered with cactus and short scrub. This was one of six or seven islands off the Californian coast. The fact that British Columbia has hundreds of islands and we have only a tenth as many boats made us realize how fortunate we are. To make sure they get a moorage spot or a buoy to tie up to, many local boaters leave early on Fridays to beat the weekend rush. With all the official moorages filled they would be left looking for a less protected spot. We dropped anchor in Pelican Bay, snucking in close to the rocks to minimise the swell; all the same it was a rough night even with our stern anchor out.

A highlight of this island was the huge Painted Cave. The entrance was so massive that we actually motored *Sky One Hundred* partly into the cave before kelp prevented further access. With a slight swell running, caution took over and we dropped anchor outside. Then with the *Restless Wind* crew, we took our dinghies in some 500-600 feet to the back of the cave. With the cave roof closing down on us the gloom steadily overpowered our combined flashlights and it got quite spooky, especially with the booming of sea on the rocks, so we returned. Later Jerry and I went again in one dinghy with the kids. We hoped to go further, under a low roof at the end and into a second cave. We had more flashlights and used them to successfully round the corner at the end. The smell of seals was overwhelming. This was supposed to be a place where seals mate. Once round the corner it was almost dark and the loud booming of the swell on the rocks was scary. There was an aura of ominousness. Suddenly there were loud honks of seals – which scared the hell out of us – followed by crashes and splashes right beside us in the blackness as they dropped off a rock ledge into the water. After a bit of an argument we tried to go further but when the swell bounced us on a rock we decided to end the venture – was it the kids or the adults who said first, *"Let's get the heck out of here"*. We rowed quickly back to light and warmth. As we returned we could see a 1' swell running down the side of the narrow cave and *Sky One Hundred* silhouetted in the entrance. The internet has many pictures of this cave.

A quick sail back to the mainland and into Marina del Rey with its spectacular entrance; this had three separate channels marked by buoys; a large centre one for sailboats sailing in and out, and two smaller marked channels for power boats leaving and entering. At the posh looking Californian Y.C., while enquiring about moorage, Heather was approached by a lady saying *"Hello honey, would you like a glass of champagne?"* While accepting the glass Heather thought this idea could catch on real quick in our yacht club. I was so astounded to see the number of huge, shiny new cars all being polished by chauffeurs that I did not think we could afford the water but it turned out to be a special car show. We were berthed between a huge power boat and a 65' sailboat. It was rather like being in a canyon and seeing the sun pass over at midday. Once again we had offers of help and the owners of the 60' power boat next to us took Jeremy and Erica out to buy ice creams. Others took us for a drive to Santa Monica and

Malibu beach. Amazingly there were restaurants, pools and sports facilities but no lift facilities for cleaning and maintaining a boat in any of the yacht clubs we had visited so far.

On September 24, 1977 we sailed down to Los Angeles, an overwhelming port area, and were fortunate enough to not experience foggy days. We stayed at the L.A. yacht club at the back of the harbour. Here we experienced some boat surge movement in a slight swell. The clubhouse and showers were very nice but, being virtually empty, the club was kind of ghostlike. Next day Jeremy and I dropped Heather and Erica off at Ports O'Call Village for a look around and then we explored the huge area outside of the coast line that had been enclosed by a very extensive breakwater; this made for excellent sailing in smooth waters. We even joined, uninvited, in a reverse handicap race for a while before sailing over to the British *Queen Mary* which dominated the area as did some small islands made to support tall drilling rigs. The old ship still looked pretty modern.

Jerry, Randi and some other sailors going offshore went with us to the French Consul in Los Angeles where we had to obtain our entry permits for French Polynesia. Here we ran into bureaucratic nonsense. On passing our application forms through the little keyhole window we were advised we could have a three-month permit which was fine, except that the period commenced as soon as we received the permit. We advised that we would not be entering the islands for at least two months as we were sailing there and not flying. The response was, that was the rule and they were not changing it in spite of our many explanations and requests. As impatience with their attitude grew we did not start a riot but we kept up a lot of protest noise and did not allow other customers to be served. After 30-40 minutes the staff eventually got the message and asked what starting date we wanted and then we all left happily with our permits. Such a simple issue really.

We sailed over to Catalina Island where we experienced some calm anchorages and enjoyed sailing by the huge volcanic cliffs. For the first time we were seeing clear water below the keel and even saw a large ray fish. Ashore we found herds of bison which we carefully edged by while they eyed us – we waited for the possible thunder of charging hooves. Somewhere up the hill Heather got separated from us and had a shock when, after some rustling in the bushes, a wild boar came rushing out snorting and complaining. He was probably as terrified as Heather. Erica also had her own experience:

> *"I had split off from the rest of the family and was stomping back to the boat when I came face to face with a bison, one of many sharing the cacti. I froze in fear for ages, realizing it would be useless running so just backed off slowly".*

We later sailed up to the Fourth of July Bay and met up with Dick and Penny, Jerry and Randi. Again anchorage was limited here to such an extent that boats did not seem to be more than ten to fifteen feet apart with bow and stern anchor lines out and very close to the rocks. It was an uncomfortable anchorage – not a place for the timid. One night after a BBQ of fish caught by Jerry and Jeremy there was a wail from *Active Light* and Penny issued an alert that the parrot they had purchased had jumped ship, or whatever parrots do when they have had enough! He'd probably run out of leather gloves. We all went ashore with them to look for the *"damned thing"* but to no avail. Dick and Penny went into a mild depression and returned to their boat – had they really grown that fond of it? Later there was a scream from their boat and they had found it brooding up in the forepeak in the darkness.

I say *"damned thing"* because two years later when they had got to Paris they called us to help them take their boat through the Canal de Bourgogne to Lyons. Well, the *"damned thing"* was still with them and its territory was the edge around the companionway hatch which it guarded with a religious fervour. It was a natural habit of any crew when using the hatchway to put one's hand on this edge when stepping down and through. This action would signal a missile strike from the parrot that would be the pride of any military group. After I was bloodied several times, I renamed it the *"green bastard"*.

Our last night was spent in Avalon, a pretty little town built up the hillside from the water rather like a Mediterranean village. Coming into the harbour I nearly put the boat on the shore. I was distracted by watching all the seals on the beach and did not realize the wind, or perhaps the current, was taking us in so close that I would not be able to sail out. Calling out to Jeremy to be ready with the anchor I went to start the engine but had to wait a valuable 20 seconds for the glow plugs to warm up. Finally it started and we were away, but it was close.

It was quite obvious here why sailors leave the mainland early Friday to get a place to moor. The whole harbour was covered with fixed buoy locations leaving no room for free anchoring except outside the breakwater. After boats picked up fixed fore and aft lines the Harbour Master collected the money; next day, he visited you sharp at 9am and one must leave in ten minutes or pay for that day. Our excessive charge of $8 was slightly offset by accidently obtaining possession of the Harbour Master's ballpoint pen. If one was rich some of the buoys could be purchased outright on a permanent basis for $25,000!

Avalon was a quaint town but very noisy. In the evening we and Penny and Dick went to the Chi Chi Bar and watched the Ali versus Shaver fight on the large-screen television. Jeremy was particularly pleased to get in at 16 when the drinking age was 21 years. A pretty crazy rule when teenagers can go to war at 18.

Sailing down this coast was quite relaxing with relatively calm seas and mild winds. Jeremy had been whittling away making a single piece of wood into a series of caged balls and chains while Erica had been artistically making leather items for horses. At some stops we might stay for 2–3 days and explore the area, run on the beach in the morning, buy more food or just laze. The latter did not happen often, no twiddling of toes on the foredeck in the sun as there was boat maintenance, laundry, collecting or sending mail, discussing homework, meeting people, gathering information about the next stops or just discussing our first big ocean trip.

At many stops we encountered an interesting device we had not seen before. Power and fish boats would have a long arm out amidships on both sides. From the end of the arm a line would suspend a "flopper stopper" (a metal fin) down in the water. Its purpose is to reduce the roll on these boats which, unlike sail boats with keels, were quite susceptible to swell and wave movement. However, in rolly anchorages we would have liked them too.

A super three-hour spinnaker reach at a steady 7-8 knots took us into Newport Beach. At some point in open sea conditions a fish boat caught up with us weaving behind from one side of us to the other. We wondered what they were doing but kept on our course. Finally they came fairly close alongside asking us, *"Where the hell are you going, why don't you make up your mind?"*

We said, *"We are going to Newport Beach, where are YOU going?"* They said *"Newport Beach"* so I said, *"Why don't you get on with it then!"* This was not received well – perhaps they had not had a good day fishing.

On entering Newport Beach and seeing the vast complex of buoyed boats, marinas, yachts, clubs and endless waterfront properties each with their own dock and boat, the first impression was MONEY. Real estate on the waterfront was expensive and still fairly expensive three or four blocks back. *Active Light* and ourselves were able to stay only one night at the Balboa Y.C. where we found the wharfinger most unpleasant. Upon phoning the Newport Y.C. they said come on over, and there we stayed for six days of which two days were on buoys with the club tender transporting us to and fro.

With a magnificent beach nearby Jeremy and I were motivated to go running in the mornings. In the afternoons, Jeremy and Erica would get down to some school work and then go over to yack with Penny, Dick, and Dave. Heather and I thought they liked to get away from their square parents. Dick and Dave continued to give Jeremy a lot of help with his math course. Dick also gave me a sextant check with his Plath sextant and found our plastic sextant had a lateral error of seven minutes. I was very impressed with the clarity and magnification of his Mercedes-like sextant. During our stay we were both preparing our boats for the ocean crossing, Dick putting a full range of steps on his mast (I had compromised with just two at the mast top), and us cleaning the underside of the hull, and varnishing teak. Why does varnish never hold up for more than two years? I also continued to add to all my notes and thoughts on building *Sky One Hundred*.

Before we left here, Dick had his 25th birthday, and Erica had made him a cake. We rowed to their boat in the darkness with sparklers lit singing "Happy Birthday". For that occasion and that only, did I wear my blazer – *"You were honoured, Dick."* 25 years old and I was 45 – nice to be young! The parrot celebrated by starting on its second glove!

After leaving the yacht club we anchored out along with *Active Light* on some municipal buoys near the harbour mouth from whence we intended to visit Disneyland the next day. While they were aboard with us for our happy hour, I asked a lady, who was rowing past, where she was going. It was her evening constitutional so we invited her aboard for a drink.

Joan was in real estate and she was very interested in what we were doing. She said nothing when we mentioned we were off to Disneyland in the morning. However, when we all dinghied ashore at 7am, there she was with her VW camper van under the trees ready to take us there. We were overwhelmed and still learning about true hospitality! We enjoyed the Pirates of the Caribbean, the 360-degree view Edison film and the horrific in-the-dark roller coaster. A very tired group returned to our boats.

In contrast to the usual hospitality, a Customs and Immigration boat came alongside early next morning. An officer boarded *Sky One Hundred* flashing his wallet identification wanting to know why we had not checked in. I told him we had been advised that it was not necessary after San Francisco. He admonished us and took all our particulars. He then asked where *Active Light* was going and I suggested he ask them. We could see the *Active Light* crew looking through their portholes with binoculars, probably wondering what was going on. He went below to inspect our boat and was about to leave when I said, "*Hold on please, I wish to have your identification data to enter into my log.*" Well, you have to slow these guys down sometimes. Then he became more reasonable, stopped acting like the "wrath of doom" and advised us they were looking for drugs. He never did stop at *Active Light*.

Other stops were at Dana Point Y.C. (where a local group invited us all to dinner) and Oceanside which finally led us to San Diego and its big naval base. We were looking forward to San Diego since it was a point of restraint in our travels due to weather conditions and we would be able to settle down for a while. November was the earliest date considered wise to leave to go south if the possibility of cyclones (*chubascos*) was to be avoided. This year they were coming later and further north. We actually ended up leaving on November 9.

The Harbour Master's office in San Diego provided a good

harbour plan. We motored in behind Protection Island and berthed at the famous San Diego Y.C. After we had tied up, a lady came over, had a chat, and then offered us her car for an afternoon or a couple of days. Apart from this frequent type of hospitality, all yacht clubs offered three nights' moorage free at every stop and generally charged $3-$6 a day thereafter. This was quite different from the formality of yacht clubs on the east coast where we sailed later; some wanted a guarantee from our club to pay for any charges left unpaid!

We stayed at the San Diego, the Western, and the Silvergate yacht clubs and anchored out at the Coronado Y.C. Jeremy loved wandering around looking at all the racing boats, many of which he had seen in magazines or had heard about. The San Diego Y.C. was our favourite; it had tennis courts which provided much needed exercise and we enjoyed their half-price Monday bar nights watching the latest football game while consuming bowls of peanuts, hot dogs, and beer served by two very leggy waitresses.

Early on in our stay we called the Grand Canyon Authority and arranged permits for our visit there. Early arrangements were necessary, unlike in 1960 when Heather and I had to drum up people to make up a mule party to go halfway down. We anchored out at Coronado Y.C. where Dick and Penny promised to look after our boat. We hired a car and drove to the Grand Canyon, stopping overnight at a camp site. We had no tent but the four of us slept under a tarp surrounded by massive RVs many with their flashing TV sets showing through the windows. How poor we felt especially when it started raining.

We took a room at the Grand Canyon Hotel where we booked our rides and an overnight stay at the canyon bottom, at Phantom Ranch. Heather and Erica rode to the bottom on mules and Jeremy and I elected to walk down and back. Erica was especially pleased to be riding. Even though this was our second trip, Heather and I once again looked back up at the vertical cliffs of the Bright Angel Trail we had just descended and found it difficult to see how a path could be made. The scenery was absolutely outstanding and very colourful. The depth was over 6000' to the Ranch.

The path down did not seem more than 6-8 feet wide before it suddenly dropped off almost vertically. The mules had a delightful way of walking on the outside edge around the bends, which was pretty scary as a rider could see many crisscrossings of the trail directly below. An American fighter jet pilot told us it was the first time he had suffered from vertigo. Some parts of the path were quite narrow, even for Jeremy and me, so on a mule it could be frightening. I wondered what the situation might be if a rider did not wish to proceed further but the mule just kept going. There was hilarious relief though, as on return, the mules seemed to develop an excess of wind. The first time there was a flatulent roaring blast from one of them, riders tried to look in other directions but after several occurrences the riders were having a hard time staying on their mules from laughing.

At the Phantom Ranch we saw a film on the canyon history and returned the next day, Jeremy and I hiking back with two stewardesses, Val and Chris. Throughout this trip the rock formations, changing colours, the thrill of the narrow winding trail and the immensity of this spectacular canyon made it a real must. Also a real must was a good supply of water as it was hot and dry. Back at the top we all retired to the nearest bar for a few drinks before we said our good byes. I was done in from the hike probably because I was not getting enough exercise on the boat. On the way back we slid in a visit to a power station and the Paloma Observatory.

Erica was especially pleased to have been riding on this side visit. She had been talking about the long passage to the Marquesas and how it would affect her with sea sickness. So this ride was very useful in diverting her thoughts and pepping up her spirits. I suspect, in the back of all our minds, the unknowns of this long voyage were on the simmer. Jeremy never mentioned it with any concern. Heather and I to date had had a very even and positive relationship on this trip even though we had never been in such a close and 24/7 environment. I do not think our feelings for each other had changed except that they had probably grown stronger and more harmonious, perhaps as being so reliant on each other. Few wives or partners would be ready to undertake such a journey particularly when their mate had not done it before; so I was really a happy person.

Back in town Dick and Penny said there had been no problems and had mail for us; some response from two Newsletters Heather had sent to our friends at home. We also received news that our lady tenant had problems with the plumbing, the electrics, and the stove. A phone call home and two good friends kindly looked after these worries which were quickly cleared up. Importantly, the monthly mint kept functioning. At the end of the year's lease she said she had never felt so relaxed as she had been when in our house and was actually getting back together with her husband. Later, we took our friends over to the Coronado, a fantastic place, for a meal.

Before leaving we stocked up on food and supplies purchased from Fed Mart – a massive store like a huge warehouse. They provided the cheapest goods and food. A six-pack of beer was $0.89 and a gallon of excellent red wine was $2. Our forward hold was packed with 15 gallons of wine and many six-packs. When the Schnetzlers from *Maradea* went to the store, they had so many arguments over which bargain to buy that they bought nothing on the first visit. But when they left for the Marquesas, their 36-foot Lapworth was really loaded down. We agreed with them that it was better to leave as soon as possible as it was a race between insolvency and loss of freeboard. We thought this older couple was very courageous to be travelling offshore and certainly in such an older wooden boat with its oversized cockpit.

At Freer's outdoor store we bought all our snorkeling gear and a Honda generator. Experience later showed that we should have bought more supplies, as Tahiti proved to be two or three times more expensive. At Pacific Marine Supply store we purchased a proper metal sextant, a Zenith short wave radio with a radio beam attachment and a Tiller Master. The Tiller Master is normally attached to a boat tiller where it can steer a boat automatically on a compass bearing, but I attached it to the wheel steering where it would be very useful for steering in calmer waters. This store was run by Steve and Tommy Flanagan, friends of the Sidneysmiths who saw us off from Vancouver. We also bought a lightweight Sabot sailing dinghy.

We were five weeks in town before the weather south looked good for us. It was an excellent time to socialize with other yachties, visit the yacht clubs, and generally relax. Jeremy and Erica were able to move along with schoolwork and receive results back from Vancouver in between meeting other kids, sightseeing, and playing frisbee on the Coronado sand beach. Erica again mentioned her worries about the big off-shore trip. There was not a lot Heather and I could do about this except to reassure her that the trip would be fine; we had come a long way without any significant problems except on the first few days down the coast adding that we did not expect the seas now would be so big. On many days the harbour was shrouded in mist which provided an atmosphere of mystery.

Prior to leaving, we visited the Mexican consul where visas and tourist cards were painlessly obtained along with a special block of forms printed in Spanish for use with the Mexican customs and officials. At the U.S. Port Captain's office we were provided with a magnificent clearance form:

CLEARANCE OF VESSEL TO A FOREIGN PORT

District of *San Diego*
Port of *San Diego*

"These are to certify all whom it both concern:
that *P. N. Hill* Master or Commander of the *Canadian Sky One Hundred* burden *14.67* Tons, or thereabouts, mounted with *no* guns, navigated with *4* Men, *Fibreglas* build, and bound for *Ensenada, Mexico* with passengers and having on board *in ballast*

MERCHANDISE AND STORES,

hath here entered and cleared his said vessel, according to the law.
Given under our hands and seals, at the Customhouse of *San Diego*, this
9th day of *November* one thousand nine hundred and seventy seven, and in the *202nd* year of the Independence of the United States of America.

Signed Acting District Director, Customs Officer.

We wondered about the "no guns" bit as *Restless Wind* had a number of guns on board. Heather and Erica were not too happy about being called "men". I thought with a form like this I might get to heaven!

Finally we moved down to the Customs dock to depart. *Maradea* and *Active Light* had already left to sail directly to the Marquesas where we hoped to meet up with them again.

We were fully stocked, had no more money, and were ready to go. We were somewhat perturbed to find the Customs and Immigration officer quite aggressively questioning why we would ever want to go to Mexico.

On starting the motor to leave the Customs dock we were even more perturbed to hear just a click from the engine. I was horrified to see the back end of the starter go red hot when I turned the key to "On" and a shower of sparks fly into the air like a mini comet. We managed to sail back to a berth at the previous yacht club we'd stayed at, paddling the last 50 yards and arriving at 8pm. I immediately took the starter off and found a space washer had got between the motor and the frame. Who knows how? After re-assembly there was still a click on starting. What the hell? I took it all off again and dissembled the solenoid that engaged the starter gear with the engine. Unbelievably two wires had to be unsoldered in order to open up the solenoid. Obviously the manufacturer wanted to make certain this could never be opened at sea – if they thought about it at all! On opening it up I found the two contact plates completely covered in carbon. Once the carbon was removed the engine started right away. Why did this happen after only 150 hours of use, when our cars run for years without such problems?

After two hours of sleep we left for Ensenada at 1.15am, November 10.

A walk
with
Antionette

Lost mast

Manihi motus

Sunset before trouble

Rain squalls

AHE, TUAMOTUS

Walking on beach with our boat in background

Coral head problem

Harbour Hopping to San Diego

PAPEETE

81

Home on the Waves

Transport styles

Tahiti from Moorea

COOK'S BAY
MOOREA

Active Light behind

Our chain
30' down

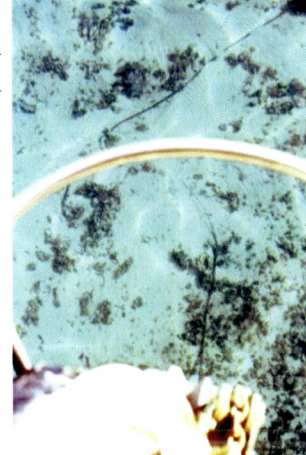

Coral
watch

Fare, Huahine

Home on the Waves

Fare, Huahine

The 'Spin'

Dick and Penny on *Active Light*

Bora Bora at far right

84

Harbour Hopping to San Diego

Sky One Hundred at Moorea

Raiatea Tahaa, a blue lagoon paradise

CHAPTER 6

Mexico

"saw a 1 1/8" diameter jet of sea water spurting in where the propeller shaft should be"

November 11 – After a wonderful starlight sail we arrived in Ensenada, our first Mexican stop, anchored close to *Restless Wind* and went aboard for a happy hour, it being too late to register our arrival with the Mexican authorities. *Snark II* was anchored nearby. The next day was to be our first experience of the Mexican three-step: visit the Port Captain's office, the Customs and Immigration office and then back to the Port Captain – all in different parts of the town. The officials were all very friendly and it certainly helped to know some Spanish. The town was rather touristy but pleasant with many bargains, especially the colourful blankets and interesting leather goods which Erica wanted to make things with.

On the way down I had wanted to take movie films of the

many dolphins seen but realised I had left the camera with a friend. I had to immediately return by bus to San Diego to collect it. The returning bus travelled at high speed and I was pleased to arrive back in one piece. While I was away Jeremy and Erica were able to try out the Sabot and Heather and Erica made cookies and cakes, Yummy! Although the Mexicans were very friendly we were ready to depart after a couple of days, leaving the roar and the swirling dust of the beginning of the 500-car Baja Race behind us. *Snark II* was going to stay a year in Mexico.

With *Restless Wind* we sailed 136 miles down to Turtle Bay or Bahia Tortuga. We heard this bay was sheltered, pretty and a must-stop. Erica was a bit sick in the night when we crossed Bahia Sebastian Vizcaino out of sight of the coast. Heather, Jeremy and I maintained 2 hours on and 4 hours off watches. Our course overnight was good as early morning we picked up Cedros Island dead ahead. The island was very bleak and uninviting but there were lovely smooth waters as we slid behind it. We called up *Restless Wind* only to find they were on the other side of the island.

We passed Pte. Eugenia on the mainland coast and started looking for the bay entrance which we expected to be small but found it to be 2 miles wide. It was poor timing to arrive in the dark and we had to carefully motor into the bay to avoid a reef. We gave advice on our VHF to Jerry about the entrance and the reef. Don't mess about in boats close to land in the dark! We stayed for a couple of days in the pleasant calm bay, sailing our Sabot dinghy, swimming and collecting abalone shells. One night we dined out with Jerry and Randi on excellent enchiladas in a house overlooking the bay while our kids joined theirs for supper. Having *Restless Wind* travel with us was working out well, especially with the children getting on together.

The little village here consisted mainly of adobe houses

and had a fine bakery. I think Jeremy and Erica were curious about the dirt roads and the general living conditions. They were certainly very impressed later to see men and women locals going on their knees up a pathway to their local church. Heather and I shook our heads thinking about the power of the priest. Earlier, when we had lived in Quebec we had heard stories that if a villager did not turn up for church he could receive a visit from the priest to find out why, and the pressure could be increased. However, we never did see the practice which existed in the Philippines on Good Fridays, where, at special religious festivals, a man would carry a large cross on his back in a parade and later have his hands nailed to the cross with a stainless steel nail for a short period of time; Catholicism is a powerful religion.

Our journey was about 900 nautical miles from Ensenada to Cabo San Lucas, 200 miles across the Gulf of California to Mazatlan and a further 300 miles to Manzanillo; a total of some 1400 miles. As we wanted to leave Manzanillo after Christmas or New Year we had 6-7 weeks travel time requiring 35 miles per day. Therefore we would have to get up early! While Jerry and I preferred to sail off-shore from Puerta Vallarta, Heather elected to leave from Manzanillo in order to see the more tropical coast further south. This turned out to be a good idea.

We had charts but they had limitations since no soundings were shown in bays and inlets and they only showed prominent high points immediately inland. They did show sand beaches and cliffs which were useful for navigation. Our charts were supplemented by a compact Baja Cruising Guide by Vern Taylor; this showed a multitude of data including navigation lights, soundings in bays and inlets, towns, supply stores and sketches of places to anchor – an excellent publication.

From Turtle Bay we made a two-night, three day 255 mile

hop to Magdalena Bay. This was across a huge indentation of the coastline where we crossed some 200 miles from one cape point to the next, out of sight of the coast, being 40 miles offshore. We caught a fish and then a bird as the lure was skittering along the wave surface. We finally managed to get the fish net over the bird and released it. Dolphins came around us en masse but were difficult to film with sunlight on the wave surface. Our shot on Polaris, after Heather had done the calculations, gave a latitude which agreed with our dead reckoning. It was quite warm now and Heather did not need to wear anything on her 'African' feet (they are wide) at night. We talked on the VHF to Jerry who was not in sight and saw a few ships in the distance from time to time.

About midday on the third day I was getting a bit concerned that I could not see the next cape point, as we had sailed the right mileage. I went below and checked my dead reckoning only to find a silly error; I had been taking a back bearing on a large mountain but found I was on the wrong bearing having used a true bearing instead of a magnetic one.

We immediately turned through 90 degrees and headed eastwards for the coast. It was shortly after this abrupt course change when a U.S. Coast Guard vessel, complete with helicopter #625, appeared over the horizon heading straight for us. As it closed in I took a movie of it through a port. This was just after the time when the Canadian environmentalist McTaggart on *Greenpeace* had been beaten by the French coast guard. So I wanted a record in case anything untoward happened. Not wanting them to stop and board us, as we had experienced in Newport Beach, I wanted to appear as innocent as possible and said to Jeremy, *"Let's go on the deck and look as if we are changing a sail."* Nothing did happen but the vessel circled around us a hundred yards off, with no radio call or sign or hand wave to us and then went off. We could not believe that in the middle of the open sea there would be no friendly wave.

It was a bit threatening but we suspected they had seen our course change 20 miles off the coast and wondered if we were drug runners. Why was the U.S. Coastguard operating in Mexican waters? Earlier Jerry and I had been told by the San Diego Coastguard not to have general chat on Channel 16, the emergency channel, but to pick another channel. This was a bit of a shock, they being 400 miles away, when VHF radios only normally operate over a line of sight distance.

After our course change we made landfall at dusk at 8–9 knots after what had been a superb sunset. It is not good to be looking for an anchorage in a 25-knot wind without a detailed chart; we only had our *Cruising Guide*. We started to anchor in 80' feet but there were rocks around and the area was rolly. We decided to cross to the north arm of the bay where it was more sheltered behind a sand bar and dropped the hook in 20' feet. I was so tired by then that I heaved our 50 lb anchor over the side as well and went to bed.

In the morning we saw several small fishing boats anchored off the beach and a number of run-down shacks on the beach. The bay was huge, about 15 x 20 miles. It was still blowing and rough so we did not go ashore and hoped the wind would not swing round to the east and come across the bay at us. Jeremy and Erica worked on their course assignments, Heather cooked and I generally tidied up. We had an early supper and hit the sack for a read and a rest. Heather was still working her way through an interesting book called 'Trinity'.

Next day, as the wind eased, we pumped up the dinghy and rowed ashore. We had a chat in 50-50 Spanish and English with a few transient fishermen. They came from La Paz between October and January, lived in the shacks and fished for lobster and sharks. One man came with his wife and three kids for three months. The marvellous sand beach was covered in rusty tins and old fish bones; not a place to walk in bare feet. While on shore a fisherman came in with three four foot sharks and one six footer. The latter had taken a neat chunk the size of a rugby ball out of one of the smaller sharks. My immediate thought was *"Don't call me for swimming"*. He had a small 15' dory type boat. It would have been interesting to see how he caught the sharks and got them into his boat. We thought it would be a good challenge if fishing tourists tried to use these boats to catch their sailfish. Later in the day *Restless Wind* came in. Over happy hour they told us they had sheltered from the wind further up the coast. We both stayed there for another couple of days.

The marine life along this Baja coast was fantastic. One day the surrounding hundred yards of sea suddenly burst into foam with dolphins leaping and plunging right up to and around our boat. It was stunning and spectacular; singly or in groups they would leap high out of the water and plunge below the surface where we could see them flash by. Jeremy and Erica were very excited and Kodak shares would rise again. Heather was not so excited when one surfaced close by with a "phish" while she was quietly dreaming away her watch! We had also seen turtles and often what seemed like a black blanket leap out of the water ahead only to realize it was a manta ray. Mantas can have huge fins up to 23' wide and can leap out of the sea completely which would be an astounding sight. In smaller shoals they could be seen leaping out one after another. We also managed to catch a few nice-sized fish which helped to vary our diet. One still had to ensure the fish were not poisonous in any way. Once, the crew on a boat anchored next to us were in bad shape from eating a fish that was not acceptable. There were still many pelicans around and for the first time we saw frigate birds and hawks. While I was ashore painting our name on the dinghy I saw some vultures.

Just north of Cabo San Lucas, around November 23, we

passed the massive and impressive *Sun Princess*. Knowing that it went to Glacier Bay, our Alaskan destination, Heather called them up on Channel 16 –

> "*Sun Princess, Sun Princess, this is the sailboat Sky One Hundred off your port bow.*"

> "*Yes, Sky One Hundred, what can we do for you?*"

Heather: "*Sun Princess, go to Channel 68 please.*" Once off the emergency Channel 16, Heather advised that we were eventually heading for Glacier Bay where they had visited and could they advise us of the key tidewater glaciers to see. The Deputy Captain, David Lumb, gave us some advice and also promised he would call Heather's parents when they arrived in Vancouver and let them know how we were doing. A year later on our trip, we and Heather's parents amazingly met the ship and David in Glacier Bay and Skagway and had a great get together on board. An ongoing friendship developed with David who, when he was in Vancouver, would invite us to the ship or we would invite him and his crew to our house for a change of scenery.

As we swung around the southern tip of this long peninsula we were excited to see rocky outcrops, a rock arch and wonderful yellow sand beaches. We were surprised to recognize the cruise ship, *Fairsea*, which we had last seen in Australia in 1963 when we were about to leave after a year of immigration there. We were lucky and managed to find an area in which to anchor in the very small harbour with some ten other sailboats and deep-sea fishing craft. It was a tight fit and we were frequently concerned we would get hit by the mainland ferry which docked there coming from Mazatlan. It was years later when several boats moored outside the harbour were lost on the beach as anchors dragged in a sudden, short (two-hour) severe squall. The famous French sailor Montessier, who we met later in our trip, also lost his boat, which was partially buried in the sand beach.

Cabo San Lucas appeared to be developing into quite a tourist spot with one or two large hotels on the Pacific-side beach and also on the mainland side. We snorkelled many times among sheltered waters around the rocks on the inland side and Jeremy surfed in pounding Pacific rollers on the outside. The water was so clear we could see his body floating vertically in a big surf wave; not for me though – a virtual non-swimmer. When the water did get warm (about 75F degrees or higher) I tried snorkelling. I started by hanging off our overboard ladder and tried just breathing properly with a mask on; then I swam to the front of the boat and hung onto the anchor chain. Finally, I would snorkel across to adjacent boats. In the end I was confident enough to snorkel anywhere.

While the sand itself was crystal clean and the water clear green, all was not well as there was an unbelievable litter of rusted tins, bottles, beer cans and garbage lying everywhere, on the beaches and in the water, ready to slice a foot open at any time. The litter was even on the gravel streets. Seemingly everyone littered; Heather asked a gringo why he threw his bottle into the harbour and the response was, *"There's no law against it, when in Rome etc. etc."*. It was a paradoxical situation because every house or shop front, be it paved or dirt, was swept immaculately every day.

The tourist industry seemed to revolve around ocean sport-fishing for tuna and dorado but mainly marlin, which were weighed at the entrance of the harbour. One night we saw a 500-pound 14' marlin being weighed; a marlin may weigh up to 1,500 pounds. It seemed criminal that these huge beautiful and impressive fish, which can attain speeds of 110 kph were being caught; the final sadness was to see them being weighed by their tail. We would rather the sport fishermen tried their luck at catching sharks as the transient locals do – in a small dory boat. Smaller fish abounded and even in this little harbour were always leaping out of the water and sometimes one or two might be found in the dinghy. The sea seemed to be alive with fish of all sizes.

Late one afternoon when Heather and I were strolling through this very relaxed and friendly settlement, the sound of music drew us into a back street. When we looked over a fence we saw colourful wedding festivities in full swing. To our surprise and pleasure we were suddenly waved at, and warmly invited in to meet the beautiful bride, the handsome groom, and to have a drink. A three-piece band played the usual vigorous music and in no time we seemed to be dancing with all the guests and children under the trees – a splendid evening. As we strolled back to our boat we could not but wonder at the natural friendliness of the people.

The bureaucratic scene continued with its ludicrous games. A permit was required to buy fuel (20 cents a gallon); on obtaining ours we motored round to an exposed high-level wharf, dropped our anchor forward, backed our boat up, and, after some difficulty, tied our stern lines onto the wharf. At this point, the service man, who had been watching us from the wharf, told us he was closed for two hours for lunch! Nice guy! We left and I returned in the dinghy and filled up a number of jerry cans.

Another form was required from Customs to say I would not sell our boat or any items on the mainland (Baja was duty free); after making four trips to the Customs to be told the Chief was out somewhere I said to heck with that – hopefully we will not be liable for a fine in Puerto Vallarta.

Two ongoing hassles on the trip were buying food and ice. I am amazed at how much food we consumed; eight loaves purchased at a time was very necessary. It seemed that when we were in the city we were always lugging food to the boat and it was a pleasant rest to be at sea again. A good rucksack or collapsible wheel trolley is a good idea as it would avoid developing the arms of a pelota player. We used a 15lb block of ice every two days in spite of covering our cooler with towels. A 12/110V Norcold top-loading fridge would be useful where every item could be frozen while on a 110V power source and the 12V system used only when the engine was running, to avoid drain on the batteries. Most importantly though and prior to leaving, we spent a day filling our tanks with water from a fresh water tap on the beach. We never treated this water and never had a problem, having been told it came from a clean artesian source.

> *"We are now 10 miles out of Cabo on our way across the Sea of Cortes to Puerto Vallarta. The wind is from the south-west at about 10 knots and the sea is flat calm — a perfect sail. Without doubt this is fine cruising coast with the medium to light conditions. Some days the air is so clear that one is deceived as to the apparent distance. At night the stars blaze out and with the moon rising early and setting shortly before dawn the horizon is visible most of the night; we can do star shots at any time. Polaris is steadily getting lower, 22.5 degrees now, and the Big Dipper only rolls into sight half way through the night. We are getting to recognize more stars and sights are working out well.*
>
> *Just had an ice-cold GAT (gin and tonic) for happy hour, while Heather sticks to BAT (Bacardi and tonic). Night falls and Erica is trying her first night-watch alone complete with her safety harness".*

December 2 – We arrived late one evening in the fast fading light, turned around Punta Mita, the northern tip of the large Bahia de Banderas and into a small inlet. We were ten miles from Puerta Vallarta located at the eastern side of the bay. Once again we were arriving at dusk but it was one of those magical moments as we motored quietly along and the sea was absolutely calm. We just dropped the anchor in 20', backed up to ensure it dug in tight and then relaxed in the cockpit absorbing the warm night and delightful peace of the anchorage. We could hear undulations of the surf on the beach and could see a few lights on shore. Soon the need for sleep crept upon us and we headed for our bunks. Later that night *Restless Wind* arrived and we heard the rattle of their chain as they dropped anchor.

In the morning light there was an air of excitement as we saw a beautiful deserted yellow beach backed by lush green trees. After a quick breakfast we pumped up the dinghy and rowed ashore with a safe landing through the surf; we were managing this much better now. Even innocent looking surfs could quickly grow large and cause chaos or worse, turn the dinghy over and us into the sea. We had to get as close as we could in behind a wave, row in fast, leap out and pull the dinghy up on the sand before the next wave arrived.

The beach was excellent, deserted, and virtually untouched. We walked up onto a higher plateau to a small village consisting of just a few fishing shacks and some poor adobe thatch-roof houses. There were a few people going about their day while chickens, pigs and goats wandered around amongst tethered horses. There was some evidence of a large development that apparently never got past the first floor. It was all quite primitive. Now it is a major tourist resort with low-rise hotels. Back at the beach Erica had found a little lagoon where there were lots of hermit crabs and birds.

Our two families spent a relaxing two days swimming and snorkelling. Jeremy tried surfing here on his board for the first time with some successes. We BBQed on the beach each night. Strangely, we had little contact with the locals. It was quite warm now and our Wind Scoop was working to perfection. It was a nylon scoop that we attached to the forward hatch and it funnelled any wind down the hatch and through the boat. It was almost too chilly to be able to stand directly under the hatch.

In Puerto Vallarta we moored at the local marina for a week. It was located four miles north of the town and just north of the docks for the ferry and tourists shops. We had the choice of anchoring out for free but chose to moor stern-to for $1 per day. It was good to be in close contact with people for a while and hear all their stories. It was particularly favourable for Jeremy and Erica to meet up with other children. It was a simple dock; the shower was a small inadequate plastic enclosure on the dock fed by a standard hose. The good thing was getting a hot shower as the hose was laid out in the sun!

Jerry and I went in our dinghy across to the Port Captain. There we encountered a frustrating two hours being shuffled from department to department. Apparently, we should have obtained in Cabo San Lucas, which was duty free, a form saying we would not sell our boat in mainland Mexico. Finally, I blew up and asked for *el jefe*, the chief, and with him we managed to settle the problem. Then we had to take a bus into town to check in with Customs and Immigration and get a form stamped to take back to the Port Captain. Hey! This was a frustrating waste of time and, unfortunately, I picked up the wrong form and had to get the right form the next day. I tried to forget we would have to repeat this whole performance when we departed.

When one can zig-zag across Mexico in a 10 x 50 foot recreational vehicle without all this rigmarole we were encountering, my mind glazed over. Where did all these copies of bumph go to, who read them, what decisions resulted, who gave a fiddle (or a fuddle) whether I had a 1.85 metre draft or two months supply of food, what dark room would they be stored in and where (Mexico City ?) and finally who has the courage to throw them all out and eliminate the whole time wasting process? I was obviously in a bad mood!

After a while we met up with Mexicans who were running parachute rides from their boat. We booked $10 rides for crew of *Restless Wind* and ourselves while receiving a small discount. They took us in their boat down to the hotel beach where two Mexicans laid out the parachute on the sand while the first sacrifice was elected for the ride, settled into the body harness and given instructions. Even little Rachelle went for a flight. It was exciting watching the shuffling run down the beach of the sacrifice supported by two Mexicans and seeing the chute rise off the beach. At this point their boat accelerated and the parachute rose suddenly 2-300 feet in the air and over the water accompanied by our cheers.

On my turn, I found the view was spectacular and, as I took pictures with my camera, I could spot black rays below in the rich blue water. The sensation of drifting along with the wind gusting past was a must, but I was not without certain and varying levels of concern, as I kept thinking, if those straps round my legs disconnect, it is a long way down. On return, when I was passed close to the beach they whistled and I had to pull on the beach-side line to stall the parachute so that it would drift over the beach as they slowed the boat even more. The intent was that I would land in the usual 20' radius. I did not get far enough over the beach but landed just in the sea. As I was landing I thought they were runningto catch me but no, they wanted to keep the chute out of the water — there were lots of spare tourists! *"Senor!, hay muchas touristas pero tienen solamente una paracaidas!"*

Years later for my 60th birthday, Heather bought me three parachute rides on Whistler Mountain. This time I laid out my chute on some avalanche slope and on the word "Go" I had to ski down the slope while the chute rose up and I was lifted some 200' into the air, on my own, floating past rock faces to land further on down the mountain. Was she trying to get rid of me?

After our thrilling rides in the sky we all strolled down the beach and into town for a stiff drink. There were tourists everywhere, shops with bargains galore and Sambo's, the ice cream parlour where the children really went to town with their choices. There was an excellent market store and a couple of hardware stores. At the market, I was intrigued to see one girl spending all her time non-stop weighing, pricing and packaging every purchase, be it 20lbs of potatoes or a couple of peppers.

One night when walking near the cruise ship dock I said to Heather and the kids, *"You go on as I am going to see if I can get on board."* Amazingly, I was able to walk up the gangway, just wave to the watch officer and go on board. There were no security checks. As I explored around I was amazed at the immense size, the luxury of the décor, the noise and the number of people. I wondered how they would react to the minute environment in which we lived and would probably be equally overwhelmed. I strolled back to the quietness and calmness of our boat.

While we had had a good sail across to the mainland Heather developed an abscess in a back tooth which became quite painful although 222's eased the pain. It needed attention before we left Mexico. By chance, in the market we were introduced, by Don off *Paloma* from the marina, to a lady doctor who gave us the name of a reliable local dentist.

We immediately went to his office where he inspected the problem. He took an X-ray and advised it really had to come out or it would get worse. This was an unfortunate decision to make, but bearing in mind we would be off-shore soon, Heather agreed. After several injections, with Heather still in a bit of pain, he managed to take it out but in two pieces while I sat there. I asked if there were any *"fragmentos"* left and he advised *"no hay fragmentos"*. Later that night a small fragment did come out. While the tooth healed well, there was no problem thereafter except Heather did develop a rotten cold which went to her chest. She was probably a bit run down. The remarkable cost for this service was $7.50.

We explored the town, ate out, visited Gringo Gulch and swam off the beach. This was not pristine like the beach at Punta Mita and we only swam there a couple of times. We enjoyed meeting up with the many yachties for chats and drinks. Jeremy and Erica were certainly pleased to be around other children for a while and even managed some school work. One day about twenty of us went to some waterfalls and pools up in the mountains for a super day with the more adventurous shooting the waterfalls. On leaving, one yachtie left all her garbage, tins and bottles under a rock. I suggested she did not want to really leave it there, but yes, she did. I once read a book where 'man' created so much pollution the surface of the sea was prevented from evaporating. The result was no rain and massive droughts. The long term effect was that populations declined along with the pollution problems until, by natural processes, the surface of the sea cleared again. Certainly deposits on bottles and tins would ease some of the visual pollution problems.

When travelling, and indeed at any time, one meets people having particular characteristics. Two types Heather refers to are "Everest" people or "ander" people. The sailor next to us in the marina was a third type, a "nadi" (natural disaster) person. In moving his dinghy he managed to knock his outboard off its support into the sea. He then spent a day locating it, hooking it out and cleaning it. With that done he attached it to his dinghy and then went for a rest below presumably believing it was a job well done. Unfortunately, the dinghy had a leak and slowly the outboard motor slid beneath the surface. When he came on deck again his language was quite colourful. He later wanted to borrow our ham radio to check his radio aerial; I took our radio over but stayed with it the whole time.

"Everest" people are another breed; on meeting, they might ask you what you have been doing and you might reply, "I recently did a record climb on Everest". Without any real sign of interest they will then enter into an interminable story of how they hiked up a local hill and got lost in the fog with their dog, or, worse, it will be their friend who went up and got lost! If we met this type at a party Heather and I would signal each other *"Everest"* as a warning to stay clear. The "ander" people are not much better; they end every sentence, comment, or even paragraph with "and er..." – a skilled technique for ensuring the more tolerant audience cannot get in a word; rather like the effect of a pre-emptive bid in bridge.

We had a visit to Yelapa south of Puerto Vallarta. Heather thought the spot was gorgeous, having a palm-fringed beach with thatched houses poking through with mountains in the background; she imagined Nuku Hiva in the Marquesas would be similar. This was an awkward and rolly anchorage. The steep drop-off from the beach required a lot of anchor chain out and we were still close to the beach. There were a couple of other sailboats along with *Restless Wind*.

After a visit to a waterfall for some swimming, horse riding for the kids (which Erica loved), and an evening BBQ on the beach, we left the next day, December 11, and made a 35 mile run to Ipala, a small sheltered bay. Both Randi and Heather were a bit out of sorts with the trots and feeling lethargic so the rest of us went on the beach for our usual BBQ. The beach was covered with fish and turtle bones left by successive years of transient fishermen who resided nearby on the beach.

Erica enjoyed exploring the great empty white-sand beaches that went on forever, tracking down hermit crabs, chasing elusive lizards through the dunes and seeing what could be found in the flotsam and jetsam.

One day, halfway through our beach BBQ in the dusk, five young fishermen strolled up and sat down around us. It was intriguing the way they just quietly arrived without interruption, as if waiting for us to recognise their presence, which of course we did. We started talking in Spanish and sign language about the safety of the anchorage, fish they had caught, girlfriends they had and so on. I had offered the first to arrive a glass of our Fed Mart burgundy which I liked; to my great surprise he took a small sip and immediately spat it out in a most emphatic manner. His mates did likewise so I assumed their tastes ran on sweeter lines. From their looks they obviously thought we had some weird tastes too. A couple of hours passed quietly talking under the starlit and tranquil night before we all shook hands and departed to our beds.

Next day we sailed off early for a 50 mile run down the coast to Bahia Chamela, another sheltered bay. We dropped anchor in 12' opposite a superb, sand beach with a couple of small cantinas, some campers, and trailers. Heather stayed in bed again while the rest of us went ashore for some frisbee exercise. Jeremy later went snorkelling with Randi and JJ.

That night we heard shouting over at *Restless Wind* and saw a fishing boat alongside. It turned out some fishermen were after whiskey and cigarettes. With no luck there they came over to us. There were four lads a bit worse for booze, one had already fallen in. They hung onto *Sky One Hundred* quietly asking over and over for hand outs.

Believing the most effective deterrent to this was silence I said not a word and would only be seen as a shadow in the darkness. I had a winch handle ready if one tried to come on the boat. With nil response they finally moved off. This was the only instance of any interference from Mexicans that we encountered – pretty innocent at that – in spite of the dire warnings by the American customs in San Diego about theft and hassling.

We crossed to two small islands finding a nice sand beach in an anchorage that was only a bit rolly. That night it was ashore again for a BBQ – we were getting quite organized by now – and spent a pleasant evening cutting up and frying abalone which Jerry had said we must try. In spite of pounding it to pieces we still found it very tough.

Sailing along with our buddy boat *Restless Wind* on an overnight trip, we came close together for a while to share a happy hour by calling across the 20-yard space. As night fell, we said, "*See you in the morning,*" and pulled away. Heather went below to make supper and immediately called out, "*The floor is wet and I can hear running water.*"

This was not the news that any boat owner or cruiser wanted to hear. I immediately dived below, found water over the top of the batteries and, on opening the engine room door, saw a $1\frac{1}{8}$" diameter jet of sea water spurting in where the propeller shaft should have been. What a shock! The shaft had slid out from the coupling again but this time had slid right out through the stuffing gland in the hull. I asked Heather to call up Jerry on the VHF and have them stand by.

In instances of crisis I think one's mind, mine in this case, tended first to be absorbing and reacting to the problem and not the solution. Perhaps my mind immediately thought of all the possible downstream problems.

However, Jeremy immediately reminded me that we could not lose the shaft and propeller because our skeg was in the way and also because our zinc protection on the shaft would not let it slide out and be lost. I stopped the leak using one of several wood plugs we carried while Jeremy and Heather quickly dropped the sails and stopped the boat. It was dusk by now but Jeremy took a snorkel, dived over and in a flash had pushed the shaft back. I was waiting in the engine room and when it came in sight I clamped the coupling back on so tightly that it never came loose again. Crisis over! Heather called up Jerry and advised all was under control again. One of the advantages of building our water and fuel tanks integrated with the hull was that not much water could collect before it would be noticed. Many production boats would not have had this advantage.

During the day we saw the first evidence of Club Med located south in the next bay as two power boats arrived with various uncovered bodies lounging around. The binoculars were at a premium that day! Twenty-five miles further south we arrived at Bahia Tenacacita having passed a barrack-like Club Med and narrowly avoiding some underwater rocks at the bay entrance. As we passed close to a 50' sailboat packed with bodies in various states of rawness we were invited to a tequila night. Erica was disgusted! I declined believing I would need too many tequilas to catch up! Heather, lying in her bunk, was still feeling lethargic and was sorry to have missed the display.

As we dropped anchor in the Tenacacita Bay we saw an astounding flash of fish. Supercharged sprays of small fish literally tearing across the water surface on their tails for about a hundred yards pursued by long needlefish bounding along on the water surface after them. A spectacular moment of action and one we were to see several times.

The *Restless Wind* crew went off up the local river to explore amongst all the bends, mangrove bushes and trees. Jeremy, Erica and I followed. Heather was still resting with an odd tiredness and aching back and limbs. Dehydration or some bug we did not know but I kept feeding her lots of water and salt tablets. Our tablets were not the enteric type (did not dissolve until beyond the stomach) and so were less likely to cause vomiting. As we rowed around bend after bend in the quietness of this river we received a sudden shock when the *Restless Wind* crew came firing out from cover in their dinghy shouting and tearing straight at us. In some small openings cut in the mangrove bushes we saw small boats suggesting fishermen lived in the back somewhere.

Back on the beach we went swimming. Jeremy tried surfing and managed to get standing on some small waves. Another BBQ that night which Heather attended. I was pleased and relieved Heather seemed better as it was a rare occasion when she felt ill. In the morning I took a stroll along the long sandy beach collecting shells and visiting the site of a new hotel on a cliff promontory. I could not see how this would be profitable as there were not enough rooms and the layout was very fiddly. Erica complained that I had not woken her for the walk.

It was December 15 and I was anxious to get to Manzanillo, 35 miles away, to spend our Christmas at Las Hadas Hotel. So we pushed on the next day towards the 14,000' smoking peak of Mount Colima.

Sailing down this coast was nearly all light wind sailing although there were some occasions when we experienced 30 knot plus winds. Since we were on a loose schedule I tried to maintain an average of 5 knots, motoring if we got too slow. While we trailed a line to catch fish we never seemed to have as much luck as *Restless Wind* who had recently caught two very colourful dorado which were very tasty. Quite frankly, unless the fish was really large the effort involved in killing, cutting up and cleaning the deck never seemed to justify the sliver that appeared on one's plate that night. To the certain dismay of gourmets, a can of beans was much easier to handle.

> *"a snake fell off the top of a brick wall flashing past me in the shower"*

Next morning, the huge Mt. Colima was clearly seen. It was still belching forth smoke and we hoped to visit it from

Manzanillo. This volcano was the most active in Mexico and in 2005 erupted with an ash plume rising three miles high. While passing the coastal airport (not marked on the chart) Jeremy said he thought the engine room seemed hot. I asked Heather to take the engine out of gear but she could not. Not wanting to step on the revolving propeller shaft I asked Heather to stop the engine. I found the transmission was hot. A transmission fluid check indicated it was down from the normal month to month level. To my horror I not only found minimal transmission fluid but much boiling salt water. I told the crew to get sailing while I drained the transmission and reloaded with new transmission fluid (I had 16 quarts on board). I took out the heat exchanger and found there was a water leak from the piping into the transmission fluid.

On arriving in Manzanillo we found *Restless Wind* anchored in front of the town. We called them up, explained our problem and that we'd like to moor alongside. We had a superb high-speed reach across the small harbour with rail down, executed a nice tack up under the bow of a destroyer (the men on board stopped to watch this operation), dropped the main, bore off on the genny, neatly came alongside *Restless Wind* and tied up. I was pleased with that action until I saw a huge moon shaped swath of oil smeared all over one side of our lovely white hull.

I immediately took the heat exchanger and rowed ashore with it and my Spanish dictionary to get it repaired, a process I was not looking forward to. Going to shore I asked a local man who was working on his boat where I could get it fixed. He stopped work and rowed ashore with me saying, *"I know the only mechanic in town who can fix that."* He, Pedro, took me in his car through a lot of backstreets to a little terraced house where the mechanic lived. He was out but I left the heat exchanger with his wife. Pedro ran a boat yard with his brother building fibreglass fishing boats.

That night as a back-up to getting it fixed (it was getting too close to Christmas for comfort as all services might close down for the week) I went aboard a Greek freighter (my secretary was married to a Greek and I thought that might help – clutching at straws) and asked to see the chief engineer. He was out but I was taken to see the Captain. He was very gracious and patient, provided me with a beer and with two of his crew we sat around talking for an hour. The outcome was that I should see the engineer tomorrow at 10 am.

Early next day Pedro and the mechanic rowed out to our boat and advised the heat exchanger could be repaired. The mechanic would repair it with pieces from an old Perkins heat exchanger. It would cost $100 US which was a shock to my system. I told them I had another price coming and I would advise them. Back on the Greek ship at 10 am I was told the engineer would not be back until 3pm. I realized I was wasting my time here. A British freighter was in and the engineer there tried to fix the exchanger but when I tried it out it still leaked. I then visited another mechanical shop but no one seemed to understand the brazing that was required. So I returned to Pedro's boatyard and he took me by car again to the mechanic's house where I left a $50 deposit advising his wife I must have it before Christmas.

Two days later Pedro and the mechanic rowed out with the repaired heat exchanger. I was pleased to see it looked a nice repair job and after installing and testing it found it worked perfectly with no leaks. I praised the mechanic's work, gave him the rest of the money and paid $10 to Pedro. That might not seem a lot but it was at a time when diesel fuel was sold for 20c a gallon. Now that this panic was over my stress level dropped and life became normal again.

The town was not as touristy as Puerto Vallarta but there were many different shops and stores although there were no mega stores except for one large fish and vegetable store. We did another big shop as we did not wish to return after leaving Las Hadas hotel. Block ice was readily available on the waterfront and could be placed right into the dinghy; the gin and tonic (GAT) situation was well in hand, thank heavens! By now *Restless Wind* had left for Las Hadas Hotel across the bay and we were about ready to go. Before we left, however, one of the warships (why does Mexico need warships?) during refuelling had dropped great quantities of fuel randomly into the harbour waters. Our dinghy and hull were covered with oil. We motored across to a new quieter port area and spent four hours clearing the oil off everything.

We stayed the night there and in the morning motored over to Las Hadas Hotel marina. Here we berthed stern-to a dock and we motored straight in dropping the bow anchor and at the last moment turning and pivoting about the line as we reversed into the dock using the bias of our left hand propeller. As we backed into the dock we threw lines to helpers who cleated them ashore. The holiday within a holiday was about to start.

Jerry told us that after they had docked, Charlie Brown and his family, who were staying in the next bay in his company's waterfront property, had invited them to share Christmas with them. Hearing that we were arriving shortly they also invited us, a very kind gesture.

The hotel had been built by the Bolivian tin king, Patino, and the work was apparently minutely supervised by his wife. She did an excellent job. Las Hadas means "The Fairies" and the whole complex conjured up scenes from a fairyland with a touch of Arabian Nights. The architecture was Moorish Mediterranean, a blaze of white forms set in the hillside amid palms and flowering bushes. No two forms seemed the same; cobbled terraced paths linked vertically with circular towers having domed roofs; spiral stairways (with echoes) were intermixed with courtyards and flowers, shrubs and wrought iron lamps. There were small swimming pools lined with decorative tile, fountains, passage ways, stairs and secluded areas. Little lookouts, including a multilevel restaurant were perched on the cliff side. Everywhere there were small scenes that delighted the eye. Each night Heather and I explored another new corner enjoying the composition and flow of this brilliant creation. Hey! Perhaps I am beginning to sound like a public relations representative.

In addition to the above, we could use the pool complete with its parrot and swim-up bar beneath the palm trees, the beach, all the restaurants, a golf course, and tennis courts. There were also various cocktail parties and a musical group that played twice daily. The big surprise one day occurred in the beach shower, when a snake fell off the top of a brick wall, flashing past me. As it hit the floor and wriggled away I do know I exited the shower faster!

We relaxed here completely enjoying the pool daily. The children had a ball. Our evening with Charlie Brown, his wife and children, and mother-in-law was a delightful mix of swimming, downing too many superb margaritas and sampling paella. All this was topped as we watched the blindfolded children trying to break a traditional piñata, a clay pot wrapped in paper and full of sweets.

For Christmas Day, all the yachties pooled their goodies on the dock for a varied lunch which included a Christmas pudding made by Heather's mum. The next day the Browns used our dinghies and went diving while we used their van and drove to Mt. Colima. We later agreed that we would have preferred to stay on the beach. Another day *Restless Wind* and *Sky One Hundred* went for a sail taking the Browns and the Carobs. The latter, Sidney and Elizabeth, were a delightful couple from London. Sidney had a bad back and did not enjoy the sail. On return Heather lent him "Stories Feet have Told". This book on reflexology advised that massaging certain parts of the body can relieve pain in other parts. Sidney reported no lessening of pain but judging by the pain he found in certain parts of Elizabeth's feet he thought she must have had testicles.

Some boats travelled with a primus stove for cooking mainly perhaps because they feared the potential liability of the explosive propane gas which we used. However, primus stoves were messy to start and use, and often oily fumes tended to coat the interior surfaces in a boat. We had heard of owners who, in exasperation, had just plain thrown their stove overboard. *Restless Wind* had had problems with their stove and after a bad evening brought their food on board and cooked on our stove. They were impressed enough to go and buy a propane stove and tanks which they used successfully for the rest of their trip.

I wanted to be gone before New Year's Eve but was persuaded by the family to stay until the New Year. The hotel's New Year fiesta was held in a cobbled courtyard laid out with tables and colour lights. We were the guests of the Carobs starting with "quickie" drinks at their apartment. The smorgasbord was sumptuous starting off with a ceramic mug being hung around our necks; it was then filled with tequila and we had to dip a slice of lemon in salt, suck it and drink our mugful – very potent. This delightful evening was somewhat spoilt by the high-decibel singing by some Piaf type blasting out songs until midnight allowing little time for dancing. She was so bad that I looked for ways to cut the lines to the speakers but finally ended up throwing a bread roll at her only after our hosts had returned to their apartment.

Prior to leaving we loaded up with stocks again from a nearby store including 100 oranges and several pineapples. The hotel kindly froze two 5-gallon water containers and on the morning we left gave us several blocks of ice. This ice lasted for ten days during which we could have GATS before having to change to GAGS (gin and Tang).

> *"I tried my credit card but it did not bend sufficiently to open the lock"*

A few days before we left Manzanillo I wrote Helen and Chris (Heather's parents) a letter to reassure them we would be alright on our first big trip. They should not worry as we were travelling with another boat and had taken many precautions and would be safe. Their concern was probably just the same as, when after a short period of time knowing Heather who was seventeen, I wanted her to hitchhike with me on the Continent for three weeks. I also suggested to Chris in a delicate way that he might consider getting a hearing aid which would make social gatherings more interesting for him.

It was my intention to leave from Manzanillo before New Year. However, a situation developed that prevented this and on reflection proved rather risky for me. Jerry and I had been doing the normal "Mexican three-step" the day before all offices closed for the holidays. We had finally arrived at the Immigration office. We came with six typed copies of exit forms ready for all possibilities and our adrenaline well under control.

Surprisingly all went quite smoothly for a while but when we tried to get an exit clearance form we were told we had to go and buy one in town. This was not right at all and Jerry and I began to feel as if we were the first boats that had done this operation. However, the secretaries had shown us the blue exit form but still we could not understand this new requirement. Again we got the chief involved but were told to see the Port Captain upstairs. Upstairs and explaining we did not wish to go buy a form he said we should see the Customs downstairs. By now I was thankful we had already paid off the taxi thinking this might only take 15 minutes.

At the end of two hours in the hot offices with Heather trying to cool me down we finally were able to go to the accounts department to pay the miserable 80 cents charge. Here we were devastated to find they were closed and would open again the Monday after New Year! Jerry and I went berserk and, with what Spanish I could muster, advised the Immigration people that we were not criminals, that we were tourists and that we were not coming back. With that announcement we stormed out of the office. *Los gringos malos!*

In the taxi going back to the boat we were still fuming about the whole situation, Mexican organization and the lack of it when, to my horror, I realized I had not picked up our passports; Heather did not have them either. I stopped the cab, leapt out, grabbed a passing cab and immediately returned to the office. It was now closed. I rattled the Immigration door in the hallway in frustration. How was I going to get my passports so we could leave directly after Christmas? They must be in the office somewhere. My mind whirled into considering how I could make an entry. I looked out of an open window and saw that if I just slid along a 2'wide, flat windowsill overhang I could gain access to the office through an open window about 6' away and look for our passports. Such was my adrenaline flow after the frustration of the afternoon that I did not fully realise the implications of what I was thinking of doing!

Looking down into the compound two floors below I thought I might be spotted by the gate guard who was leaning back in his chair but getting up now and then to check cars entering and leaving. This option seemed too risky.

However, back at the door again, which I shook again in continued frustration, I found there was quite a large loose fitting between the lock and the door frame. With my nervous system at full blast and hearing no one around I tried my credit card but it did not bend sufficiently to open the lock.

Damm! but my more flexible driver's license did. In a flash I was inside with the door shut and wondering where to look.

I quickly searched the desks. In one there were some girlie pictures but no time for that! I was just opening another desk when there were steps in the corridor! I dropped down behind the desk. If found, I was going to fake some sort of medical problem. In any case I was not far off having one as my heart was doing overtime and sweat was pouring off me! Finally, in another desk, there were our passports in a plastic bag. With no sound in the corridor I carefully opened the door and was gone.

Back at our boat while downing a series of GATS in a recovery mode other yachties regaled me with horror stories of Mexican jails. They also advised that without an exit form from Mexico, it was most unlikely the French authorities would let us into French Polynesia without a big problem. Over the next few days this began to concern me so much that I decided to wait over the holiday and **return** the passports before the office opened again.

Thus, early, very early, Monday morning I was back at the office but why was the door open so early in the day? Oh, heck! A cleaning lady was there and when I entered to do my deed, she advised the office was closed and to come back at 9am. Now how I did the following I cannot remember but, with a great flourish, I entered the office crying *"Ah! La bella vista!"* while somehow sweeping her towards the open window with its view across the harbour, at the same time dropping the passports into the drawer of the secretary's desk; a performance that would easily justify an Oscar. When I returned again at 9am and the office was open, the secretary said with a big smile, *"Senor 'ill, sus passeportes!"* but she didn't know why I was smiling.

CHAPTER 7

Three Weeks at Sea

"Virtually one of the few places in the world where one does not see a strange face for weeks"

Manzanillo to the Marquesas Islands was to be our first test of crossing a big ocean – 2,800 nautical miles (5,500 kilometres) or the distance from Vancouver to Quebec City. I had a few "simple" misgivings roving around in the back of my mind, like *"Would I be able to navigate to a smallish island in the middle of a large ocean, what if we could not find it, what if we fell sick"*, and several other what ifs! I did know for certain that we would not be spending any money during that time. It took us 20 days at an average speed of 5.9 knots.

Sometimes we would sit on the shore looking out to the far horizon contemplating our trip. Our journey so far had been parallel to coasts but now we were looking to head out to the open sea for some three weeks. Days of seeing nothing but 360 degrees of horizon was a concept we were going to have to get used to. An exciting and new prospect, yes, but nevertheless one tinged with a slight slither of concern. We wondered how the early explorers felt, who sailed literally into the unknown without the benefit of the accumulated knowledge we had.

Heather had her own pre-trip views, dated December 22, 1977:

THOUGHTS ON A
FIRST OCEAN PASSAGE

With 7 months old Jeremy tethered to the aft deck rail we lay peacefully in our deck-chairs, sailing smoothly into the South Pacific aboard the luxury P&O liner, Oriana. Little did I realize that sixteen years later I would be contemplating crossing that ocean again in a much smaller boat. In about a week's time we shall be leaving Manzanillo, Mexico for Baie Tahauku in the Marquesan Islands of French Polynesia.

What is it like to cross a large chunk of ocean in a small vessel and not see land or even another vessel for perhaps four weeks? Of all the accounts I have read, no one seems to have any fear, but I must confess the thought of all those miles of ocean gives me the "heebie-jeebies" every now and again.

While recovering from a short bout of Gringo fever, I have had a chance to study "Ocean Passages for the World" and associated charts of wind directions and speed and I calculate that we have about 2,750 miles to sail — that's more than we have travelled so far and we have been going for 5 months! HELP, I can't possibly have enough food. "Stop panicking," I say to myself; if we only travel 100 miles a day it will only take about 28 days and I'm sure we will average better than that in the Trade Winds, IF we can get through the Doldrums without too much delay.

TRADE WINDS – I wonder what they are really like? From what I remember they blew consistently and with quite a velocity. Ocean Passages says Force 4 which, with a quick reference to that delightful bedtime reader – Adlard Coles' "Heavy Weather Sailing", leaves me very confused. Force 4 on the Beaufort Scale means winds up to 17 knots. However, the symbols of 2 feathers (Force 4) on nautical charts indicates 18–22 knots. Oh well, 22 knots will keep us moving along nicely.

DOLDRUMS – what an ominous sound that word has; it makes me think of facing the washing up on a Sunday morning after a good "bash" on Saturday night. From what I read of them, we will put on our 36hp motor and burn up our 20c/gallon Mexican diesel fuel and shoot through them as fast as possible, hoping to catch some nice fresh water en route in the occasional squalls.

FOOD – that four-letter word keeps rearing its ugly head and I see Oliver Twist-like crew members holding out their plastic dishes for MORE. I had better have a good final stock-up in Manzanillo and buy some treats for that last week of the passage when the crew look as if they are about to mutiny. I have become quite proficient in sprouting mungo beans to pop inside lunch time sandwiches for an extra vitamin supplement – sure wish I could grow fresh milk!

I also had a chance to read Eastman's "Advanced First Aid Afloat", an excellent book for offshore cruisers. It has really put the wind up my sails with all the detailed catastrophes that could happen to us at sea. Heat stroke and heat exhaustion look quite common (and serious) so I'll have to stock up on some more salt tablets.

I bet you're saying "She doesn't seem very well organized". But you see I have always done everything at the very last moment – a great procrastinator – it somehow seems to work and is always more exciting (for me).

Naturally, I think of the worst that could happen to us and that is to be sunk in mid ocean, so we have an Emergency Bag at the bottom of the companionway into which I am adding "things" continually. We have all read Donald Robertson's account of their sinking in "Survive the Savage Sea" and have already met one chap who was holed by a whale 750 miles out of Acapulco. He said to make as much noise as possible while whales are around – turn on the motor, rattle winches, etc. to frighten them away. We wish we had purchased more of those garish lures from the Army & Navy which the fish find so attractive down here.

Will I be able to relax and sleep off watch – something I did not seem to accomplish too well on the way to San Francisco? I expect that after at least a week, nature will take over.

Every once in a while I get carried away thinking of our first landfall. Who will be on watch, who will shout those welcome words "Land Ho"? Will there be anyone in the anchorage we know? We must make a good job of anchoring – no shouting, all hand signals – remember crew, all eyes will be on us.

I wonder what our priorities will be on arrival – cold beer, new faces, mail from home or just a peaceful night's sleep?

I'll tell you next time.

Restless Wind and ourselves departed Manzanillo on Monday, January 2, 1978, after photographing and wishing each other luck. We had the thought that we might be sailing in sight of one another for a while but by the next morning they were not in sight.

We had firm winds nearly all the way, encountered no Doldrums and generally wondered where the languid sailing in the South Seas idea came from. One boat we met took 34 days with 9 of them in the Doldrums. They did not attempt to motor whereas we would bang the motor on directly we dropped below 3-4 knots in rough water and 2-3 knots in calm water. There seemed to be no point in arriving in port with full tanks having lost valuable time drifting at sea. I have never been a supporter of those who would sail without motors and then want a tow into a harbour.

With winds from the south and on our magnetic bearing of 210 degrees, we found ourselves having to beat and close reach to maintain our rhumb line (a straight line course to destination). We called up *Restless Wind* behind us, suggesting they head south as much as possible, as we had doubts at times, perhaps foolishly, that we would not get enough southing to hit the Marquesas. This would mean we might have to tack our way southwards! We did find disconcerting differences between our dead reckoning and our celestial positions of up to 15-20 miles. *Maradea*, who had crossed earlier, advised us by ham radio they experienced a similar difference. The Pilot Chart did indicate for our route currents to a maximum of 35 miles per day. All our calculations were worked out in the cockpit where it was more pleasant and so the area under the dodger on one side of the hatchway held all our gear for working a shot. The sextant was kept in its screwed-down box in the head.

It was a bumpy trip and we were all feeling a bit queasy with the movement and getting our sea legs back, mainly for the first few days while we moved along at a steady 6–7 knots. Jeremy found the most secure place in the boat was the head and the best place to read. To slightly relieve the motion we would run off a little during the day, certainly while Heather cooked and served dinner. We would then head up more overnight while we were asleep.

Initially we found there was no way Jeremy and Erica could do their school work. For Erica the long forgotten sea sickness was back for 4–5 days but not as bad as the first time. I was also sick the first couple of days and we all felt a bit nauseous, even just reading a book lying down produced some nausea, although we did this mostly between eating, sleeping, and navigation. However, we were all able to attend "happy hour" and consume a drink or two with nuts and goodies.

When we left Manzanillo we had packed blocks of ice, provided by the hotel, into every space we could; we even had some fixed down in the cockpit. Thus we were able to have iced gin and tonics, GATS, for nearly ten days. Thereafter GATS without ice were not acceptable but we found that warm gin and orange Tang, GAGS, were just about O.K. Some months after we had returned home Heather and I tried a GAG one Sunday afternoon – it was revolting! In all our sailing Heather always produced a "daily bread" for each of us – it was a small bar of chocolate or some other goodie. Sometimes they were swapped among us for certain favours or perhaps held until the night watch or even the next day.

Several times we had the company of dolphins. One evening a booby bird landed on the transom and decided to stay the night. He tucked his head under his wing and balanced, teetering back and forth on his little legs. Obviously finding conditions inferior, he left the next day. We also started to see flying fish, a few of which landed on the boat but none big enough for the frying pan.

After a couple of days out we came into some squally weather with great displays of lightning and black clouds racing across the sky. Once a water spout appeared to be forming and, not knowing what it would be like, we dropped our sails in case it approached but it seemed to dissolve. We had stronger winds accompanied by a few showers but not enough to collect spare water. Being alone with the full impact of lightning flashes and the ripping crash of thunder, especially if there was a strike into the sea, was quite scary at times. I was glad I had installed strips of copper plating from the bottom of our mast to the keel bolts. This would hopefully transmit any electrical charge straight through the mast and keel into the sea without significant damage to electrical equipment and the interior of the boat. We had heard of a boat, when directly struck by lightning, that had most of its electrical equipment badly damaged.

We were thankful too that we installed a sun awning before leaving San Diego as the sun was extremely hot and it was very pleasant in the cockpit in the shade. What we should have had was a small square of canvas to hang up to collect rain into a water can. Oddly enough the weather seemed cooler in the few degrees from 4 degrees North to 4 degrees South; on night watch Heather still needed a jacket over her "itsy" and a light sleeping bag over her at night.

> *"Initially coming out of a deep sleep to go on watch was always a grind"*

Our key contact with the world and other yachties was our ham radio. It provided a certain level of entertainment into our enclosed environment on a regular basis. I did not have a radio license to operate it as the need to learn morse code and all the ins and outs of the machine and its operation was too much of a distraction prior to the trip. Therefore I did not have a recognized official call signal. When I first bought it I tried, at the advice of others, to use a Puerto Rican call number. But while on the air I was told that I had a false number and to get off the air; regular licensed ham operators are very possessive of the wavelengths they are officially allowed to use.

A few days later I tried to use a Costa Rican number; this time an operator kindly told me I was sounding illegal and suggested I get myself a Canadian number and call him back a few days later. Erica was reading our book of Vancouver call signs, (what a sensible girl), and said there were some out-of-date numbers, why not use one of those? This was an excellent idea. We did use this call sign, VE7XT for over a year although we found later that the number was current and I always waited to hear somebody say, "I am the real VE7XT, why are you sleeping in my bed?"

The radio was a marvellous and powerful asset for communication. Our main contacts were to other yachties and the "DDD" net for offshore yachties. This net was a wonderful support system and run out of Victoria by Jerry Anscombe and Ralph of North Vancouver, two blind gentlemen. I could not imagine how they did that but we were very thankful they did as it was so helpful. We would contact them every two or three days. It was always exciting and very impressive to be able to tune into the agreed wavelength, start calling their call signs and hear one of them reply. We would always give them our position and our weather and they would give us any hot news. We would ask to talk to Heather's parents and they would patch us through and in no time we would hear their phone ringing and her mother answering. A magnificent system and all free. Once, they put me through to my office when I had a chat with my working buddies – on ringing off I did remind them not to charge the time to my past projects. It was also very reassuring to make the contact and to hear other boats calling in at the pre-determined time. The DDD stood for the "Dreamers", the "Doers", and the "Doners"; we were in the middle stage.

Calls could come from anywhere: such as from the man in Panama who wanted me to tell him how to build a boat;

the man in England who was amazed that we were coming in so clearly right into his living room; a boat which asked us to make a check on the navigation calculations as their readings were not making sense; and, a man who wanted to send us a certificate for crossing the equator. Some would ask technical questions about our radio to which I would make evasive answers as I did not wish to be recognized as being illegal – what did I know technically about "heaters" when asked on the radio – it was hot enough on the boat as it was. Some yachties who over-used their radio or were slack with procedure were often recognized as being illegal and were promptly told to get off the air! We used our radio throughout our year's trip.

Heather always managed to produce breakfasts of oranges, fried eggs (we had 12 dozen on board all smeared in Vaseline to preserve them) and bacon. We had a large stock of bananas hanging on the boom but they soon got mushy and were used to make bread. The oranges lasted well and were quite delicious. We were also stocked up with potatoes, cabbage, carrots, onions, green tomatoes, peppers, cucumbers and a large vegetable resembling a turnip in appearance – very crunchy and nice in a salad. We also had an abundance of limes – Heather wanted no scurvy on board. Salads at lunch and cooked dinners of ravioli, spaghetti and meat sauce, meats stews, fruit, along with baked bread, cakes and cookies kept us going all day. Heather would bake bread every other day. Her speciality was salt water bread. I was always amazed how much we enjoyed bread and how quickly we could consume a loaf.

Erica also got round to baking a couple of cakes. This effort was always appreciated as nothing is easy to do when the boat is in motion. Of course, the end product was very much welcomed and was also consumed with considerable relish and speed.

Erica said, "Food was a big thing. I was such a fussy eater that it must have been very difficult and frustrating for Mum to feed me. And I don't know how she managed without a fridge. I loved her cornbread made with seawater. I can remember her sieving the flour to get the weevils out and the fact that it didn't really bother me."

Jeremy and Erica did the dishes. As one washed with sea water the other wiped and dried them. When arguments developed over whether plates were properly washed or dried we had them work separately on alternate nights.

For safety we always wore life harnesses at night and with any rough seas during the day. We had them clipped on at the top of the companionway so that a harness could be put on *prior* to leaving the safety of the cabin. We ran a safety line forward along the length of the boat to which our harness could be attached when going forward. Nobody could go forward at night without another being in the cockpit. I did break this rule from time to time on calmer conditions or when all were asleep. As another safety feature I kept the spinnaker pole attached to the mast about 5 feet off the deck and attached at deck level in the bow. This gave extra support when working on the foredeck.

It was highly important to not get sloppy or over confident in moving about the deck at any time. To fall overboard especially without a life vest, which we did not wear unless it was rough, would be very dangerous; when a bag of garbage was thrown over from time to time it would disappear from sight after passing over a couple of waves.

Our man-overboard pole and life ring could be thrown over to mark the spot but the action would have to be really quick to be effective. The thought of ever swimming in the doldrums, which we did not experience, would have no interest for me particularly as I am not a swimmer. Taking care we never lost any equipment was important too. Thus our dinghy, when in use, was always tied to the side of the boat with two lines, tools had a safety cord on them, and drying clothes were attached with pegs and a line. Once Heather asked Jeremy to "throw over" (to her!) a towel which was loose on a life line and he picked it up, whirled around his head and threw it overboard – he must have been thinking about his girlfriend – we didn't let him off about that for a long time.

For watches Heather, Jeremy and I generally went 2 hours on and 4 hours off through the night. Whether this was really necessary I don't know because we never saw any other vessels or a 'plane until the last day when we saw two ships. I preferred the dawn watch and liked to watch the sun's rays spreading over the sky accompanied with the sense of another day.

Although Erica never did night watches she *"did love being out at night and listening to the waves when, if I was lucky, I could hear the squeaking of the dolphins as they surfed and dived alongside. There were so many stars and the occasional satellites to be seen on a clear night, changing as we crossed the equator, especially recognizing the Southern Cross for the first time."*

The main function on watch was to check the correct course was being maintained. If the wind veered then the boat's course would change by the same amount and an adjustment would be required to the windvane, Goldfinger, to bring her back on course again. Checking the compass bearing would require getting up to look at it aft of the steering wheel and I sometimes wished I had put a compass in where we sat which would have been more convenient. An occasional look forward was required for any passing ships of which there were only a couple in the whole trip. We always had a concern we might hit a semi-submerged container which had fallen off a ship or even a whale but this never happened. Cruising friends once hit a whale and, as it dived, its fluke smashed down on the stern flattening the aft lifelines; fortunately it was a centre cockpit boat and the wife on watch was not hurt. A final duty was viewing the wind speed and waves and deciding whether a sail change may be required. The rest of the time was generally spent making one's self comfortable against the boat's movement, wondering where all your posterior fat had gone and contemplating the sea, the sky and life. On a clear night one could write notes or even read.

At the end of a watch we would wake the next watch person a few minutes before time in order for them to get ready. I used to wake Jeremy quite a few minutes early because he was quite a slow mover in the middle of the night. I used to watch his murky form in the gloom as he put on a sweater, and finding it was the wrong way round, I'd know that there would be more delay while he made the adjustment. He would usually listen to the short wave radio or read. He would scan the entire band range in search of a good music station. Sometimes the station would fade or he might get a Vancouver station. It was weird, as were my experiences of finding stations in foreign languages quite clear while those in English would keep fading in and out.

Initially, coming out of a deep sleep to go on watch was always a grind. When the stars or moon were visible the scene was so impressive one soon forgot the inconvenience. The global massiveness of the sky filled with stars of all sizes and the brilliance unblemished by any extraneous light was extraordinarily impressive. After a number of nights we were recognizing star patterns and individual stars. Time passed quickly watching the stars revolve and the many streaming flashes of falling meteorites, much brighter than when seen on land. The grandeur of the heavens became especially evident when coming on watch again four hours later to see where the constellations had rolled round to or, more correctly, where the earth had turned to.

Sometimes we would see a satellite zooming along its set

course; it was strange to realise it was only just a few miles off the surface of our planet while the stars were light years away.

The clarity of the heavens in the clear unpolluted atmosphere was always breathtaking; it reminded Heather and me of the times we slept out in the outback of Australia. When the moon would rise out of the sea there could be an immediate thought that it was a cruise ship because there was so much light. Most times a shimmer of light lit the waves but when it was cloudy and black one just felt the action of the waves, their direction and the motion of *Sky One Hundred*. Meanwhile Goldfinger, naturally with the help of Paddington, our good-hearted mascot, was doing its steady job keeping us rolling along on course, leaving us to make the odd check on our compass bearing and to realign Goldfinger if the wind veered.

Prior to night falling and as a precaution I would normally hank on and have ready to use a smaller sail in case the wind increased. To not be ready for a change of weather conditions at night would require more time dealing with the sail change on the foredeck – not a wise situation. For the inner man on watch there were granola bars, chewing gum and oranges. Heather's method of eating oranges was the least messy I know; she made a hole about the size of a quarter and sucked out all the juice and then tore it apart and ate the solid part. During the day we kept a general watch since we were all in the cockpit. Here we sometimes clashed with definitions, one saying to the other we were *"too high"*, referring to the wind, while the other denied this referring to the compass course. Being at sea, as we were, had to be totally unique as it was virtually one of the few places in the world where one did not see a strange face for weeks and neither did we spend any money!

We never slept in our proper bunks because it was too

bumpy forward and too much yawing in the aft cabin. I think we all preferred being together; perhaps because of some feeling of security – it is a lonely place out there. One person slept in the 7' long couch set in the curve of the hull side, one on the floor beside it while Erica slept under the table. A rather messy arrangement but the central and low location certainly had the least motion. Sometimes on the calmer nights Heather and I would slip away to the aft cabin and to heck with the watch. Calm nights were always welcome!

> *"The crossing is not as smooth as we had often heard about – gentle seas and swimming off the boat. Winds have been 25 knots gusting to 35 making the motion quite uncomfortable at times. The body systems seem to get used to the boat motions and even Erica has relaxed. However, the subject of horse trips does come up at some of the happy hours. We do know we are losing weight as evidenced by needing a cushion to sit on in the cockpit; the two lower pelvis bones seem to want to make direct contact with the hard seats! We all enjoy the "happy hour" when the sun is closing in on the horizon. It is a reflective time sitting in the cockpit with our favourite drinks and a can of nuts discussing the books we are reading, the chickens Erica wants us to have when we return, homework, cold drinks and ice-cream, whether Vancouver has changed, letters received, what we want to do on return, and even modifications that could be made to the next boat! As dusk surrounds us the diminishing disc of the sun finally slips from sight below the horizon, which disappears as quickly as darkness arrives."*

As our bodies attuned to the flux and restlessness of the sea our daily rhythm developed.

On waking, the freshness and coolness of the dawn air was quickly replaced by increasing warmth and humidity along with the appreciation it was not conducive to physical work.

Over breakfast, often taken at one's own pace, we would consider the weather, check for any forecasts, consider the tasks for day or maybe just moon around.

We all did a lot of reading. Erica *"read masses. When I had finished everything I had, I would read Jeremy's books, then Mum and Dad's books and the medical book, with all its horrific pictures."* Books were traded with other boats. Jeremy would slug away at his Grade 11 course which was quite difficult. In Erica's Grade 7 science course she was required to grow mould on some bread but, in the humid atmosphere, mould grew anywhere but the bread! Letters were written and letters received were re-read and replies prepared for the next stop.

I would check the engine, the oil and transmission fluid and run it for half an hour to keep the batteries fully charged, a very important requirement. If they got too low we had our Honda generator which we assumed could produce enough power to charge the batteries enough to start the engine. Fortunately we never had to put this to the test. Depending on the next scheduled radio call on the DDD net, we would have our thoughts and location ready for Jerry in Vancouver. Sometimes on a call a third party might come on our wavelength wanting to talk. I might respond saying *"Go, the breaker"* or *"Hold, the breaker"* suggesting he talk now or wait till I've finished when I will call him.

Heather would be considering our next meal, writing her Newsletter or reading. I would check around the deck and fittings for anything untoward such as worn or loose lines, or worse, a hair crack developing in a shroud fitting. I had a spare cable and end fittings just in case.

Around midday we would take our sun shot (difficult on rough days); sometimes I would take two shots. The moment I obtained the angle between the lower limb of the sun and horizon with the sextant, I would call *"Now"* and

Heather would start our stop watch. We could then calibrate the seconds to GMT on our Timex watch or our ham radio. With Greenwich Mean Time we could start our calculations, which Heather loved to do. Once she had worked out our new figures they would be checked with yesterday's figures to confirm the trends were consistent. If they were not it was most likely a mistake in the calculation which would then be rechecked. She would then plot a position line on a navigation pad and, by dead reckoning of distance and bearing from the last position, intersect this line and that would be our new latitude and longitude. It was drawn up on one 8 x11 inch sheet of paper. Sometimes I would add in a moon shot or shoot three stars which, by triangulation, would give us a good check on our navigation.

Every couple of days we plotted these position points on a main ocean chart that showed Mexico and the Marquesas. Slowly these small position points with dates eased their way across the chart towards our destination. Whether it was for some psychological reason or because we had done around half the trip we realized one day our phrasing had changed from "miles done" to "miles to go".

Today, GPS, Global Positioning System, makes all that work obsolete by simply reading off the screen the latitude, longitude, distance gone, distance to go, speed, position and a multitude of other data, generally including when to blow your nose. It would be best to have two GPS's onboard as we have later found. It would be wise to still have a sextant and tables just in case all electrics failed say, due a serious lightning strike damaging the boat.

Crossing the equator required the traditional attention by Father Neptune to the crew, Jeremy and Erica, who had not done this. Heather and I had in earlier years when we had immigrated to Australia for a period. As I appeared from the cockpit in a striped sheet, with my Father Neptune hat and trident, all prepared while on various watches, there was a sense of utter disbelief on their faces. *"Dad's gone off the rails."* I made the usual incantations to the N, S, W and E winds, and welcomed them both to the southern hemisphere and went through the blessing rites. Then I lathered up Jeremy's face and while he laughed when I poured water over him, Erica did not go for this treatment at all and went into serious high dudgeon for the rest of the day until happy hour when her good spirits returned. Heather and I consumed a third of a bottle of champagne each (still cool) and the kids had the rest.

Some typical non-technical entries in the ship's log:

JAN 2

1145   Depart Manzanillo
2200   Dolphins alongside with phosphorescent trails

JAN 10

0800   Lots of lightning
0905   Some blue spots in sky – wind picking up now
1000   Black clouds behind us – very stormy looking
1100   Wing and wing – doing 9 knots
1900   Sighted loads of dolphins some leaping 15 feet high

JAN 17

0300   Patrick off watch – on comes beautiful (*Heather, that is*) – overcast (*sky that is!*)
0400   Pounding along
1310   Make contact with Schnetzlers in Hiva Oa. They took 32 days San Diego to Marquesas. *Active Light* – took 24 days

JAN 14

0100   Windvane working well

0500   Barrelling along
1000   H makes bread and banana loaf – sea calmer

JAN 16

1237   Talk to Malcolm Wilkinson and get DDD patch to Helen (H's Mum)
1420   Perform Equator crossing Neptune ceremony
1800   Coffee & cake for supper.
Restless Wind 350 miles behind – following our beer cans!

JAN 17

1100   Take sun shot
2030   Windvane breaks

JAN 18

0500   Lights of ship seen to port
0900   Repair windvane
1400   H can't sleep. J and P doing 1 hr watches. Don't think wind below 20 knots for last 10 days and course is 30 degrees to waves – rolly and somewhat jerky motion

JAN 21

1320   Take sunshot
1810   Take another sunshot – reckon 126 miles to go
2400   Get Erica up to see dolphins for an hour. Clear sky. With a 100 miles to go this is a moment of truth

JAN 22

1500   Land sighted – little wind, motoring
2045   Drop anchor in Baie Tahauku, Hiva Oa. Skins all round (high fives)
Average trip speed 5.9 knots

We were running into quite a few squalls now. They were often seen marching across the horizon in separate grey blocks of cloud descending from a continuous grey to blackish overhead mass. Sometimes they would miss us but when they sweep over us it was with considerable force accompanied with an intense pelting downpour. Half an hour or so later they could be gone leaving us with a nice clean boat and often more fresh water. Many times as the squall approached we would douse the genny so that we did not get laid over or the sail damaged.

Goldfinger was doing a splendid job because, without it, the trip would be very tiring or likely impossible. It seemed to work well since we always had the main rudder dead in line with the keel. One day though, when we were quartering with fairly sharp movements through a series of short waves for several hours at around 8-9 knots, there was a crack from behind us and Goldfinger lurched sideways. "*Oh shoot!*" was my immediate reaction as I saw our aluminum angle strut had broken. We disengaged the vane from the auxiliary rudder, pulled the latter up straight and tied it in position.

We now had to steer the boat manually by the main rudder which was a big chore. It was a devastating moment and shock to the system because the first thing I thought of was the trouble we would have manually steering the boat for the rest of the trip. I did nothing further as it was almost night.

Next morning I awoke in a more positive frame of mind ready to fix it. I noticed the strut, which I had earlier found in some scrapyard, had a small drill hole at the point it had broken. Seeing the texture of the metal at the break was smooth I was sure the breakage had resulted from fatigue and not from an overload which would have shown a rough surface. The forces that came off the auxiliary rudder were always alternating the load in the strut from compression to tension to compression. This caused the strut to fail typically in fatigue and right at the drill hole. In two hours I had it repaired using spare aluminum strips I had stored and using our Honda generator to power our 110V drill. Now we were back in the groove again and it never failed again.

More marine life was appearing and often at night there was a "phish" alongside and I could see the phosphorescent trail of a dolphin as it dived under the boat. Birds were appearing and now and then a piece of brush or kelp. We were nearing the end of this passage and our excitement was increasing. At the same time though, my concern was growing as to whether our navigation was going to be correct. After all we only had some twenty sheets of paper for plots for the 20 days of travel. If we did not see Hiva Oa where would we be? It was a feeling of concern which kept growing while I tried to ignore it.

Our lookout sharpened. Clouds generally form over land elevations. We were all looking for clouds that did not move and we would know that was a peak on land or a fixed cloud over land. There were many times when we said *"That's got to be it"* but the cloud did not solidify into a land form. Finally Jeremy saw a thin line on the horizon at 1100 hours on January 22/78, our 20th day at sea, which did not move. We were some 35 miles away.

Great, we had made it, no more plotting for a month. With little wind alternating with numerous squalls we steadily headed landward. I, and I think Heather, experienced a deep sense of relief and of achievement.

While explorers had been making this crossing since the 1500s, it may seem simple now to cross the ocean, but it was a long and lonely journey and a totally new experience. In fact we found it was an environment of which most people have little understanding. Perhaps, bound by the usual physical restraints we live under, it was difficult for people to have a concept or notion of the size of an ocean where there were no markers or fixed points. Being asked on many occasions *"Where do you anchor at night?"* or even, *"Can you stop at a marina?"* left us realizing that many people, if not most, do not fully appreciate there is nothing out there except a vast, deep sea and a limitless horizon which one travels to day after day until land appears while living with the expectation that it will appear!

Jeremy and Erica "spruced" themselves and the boat up for arrival. We motored along the southern tip of the island passing giant volcanic cliffs with jagged peaks behind. For Erica, the first close views of the Marquesas were quite ominous being very dark and foreboding. It was not like she imagined with white beaches but black sand with lush thick forests behind. For me it was plain exciting.

Soon we picked up the intoxicating and exotic fragrance of the island but it was another eight hours and almost dark before we picked up the sector light of Atuona, and turned into Baie Tahauku on the island of Hiva Oa, the resting places of Paul Gauguin and Jacques Brel. We dropped anchor in 30' of tranquil waters opposite a small island copra boat at 2245 having motored in under a full moon. It was a very sweet moment. We did skins all round and relaxed in the warm fragrant airs of the soft night while marvelling at the moonlit peaks around us. We had done it and a special feeling of relaxation eased its way into my system.

We awoke the next morning to find ourselves surrounded by exciting-looking lush, green, and craggy peaks.

CHAPTER 8

French Polynesia

"We were given tomatoes, papayas, bananas, breadfruit, limes and mangoes"

Marquesas........
The Marquesas are a group of twelve volcanic islands. They are one of the most remote groups of islands in the world and lie 850 miles northeast of Tahiti. The islands with deep valleys are quite mountainous, the highest point being 4,035' on Ua Pou. Except at high altitudes precipitation is low and droughts are frequent. There are no protecting fringe reefs. It is thought that the people were Tongan and Samoan who arrived around 100 A.D. In the 16th century, it is believed 100,000 people lived on the islands. Due to Western diseases the population dwindled to some 20,000 by the middle of the 19th century, was 2,000 at the beginning of the 1900s, and is now just over 8,000. In 1595 a Spanish explorer named the

islands, later the United States claimed them as the Washington Group, but in 1842 France claimed them under the control of French Polynesia.

We found it very hot ashore as we walked a mile up to the small village of Atuona to report our arrival to the French gendarmerie and explore the local scene. It was a long walk when carrying our purchases but sometimes over our visits we would be offered a lift. The gendarme wanted many forms filled, wanted our deposit for being in French Polynesia and retained our passports until we left the island. There was an excellent bakery, where orders had to be placed the day before, and a scattering of smaller shops with limited supplies.

We stayed for five days at anchorage. While the water was warm it was murky and did not encourage swimming. It was rather rolly and dinghy landings had to be made at the west end of the beach avoiding the effects of the swell. We were pleased to find a water supply and shower there. The bay was a bit of a disappointment as we were anticipating yellow sand beaches rather than black volcanic stones, frequently littered with rusted tins and bottles.

There was no doubt that initially we were on a 'high' having arrived at this distant island, the first in the South Seas, and seeing another way of life. Strangely though we seemed to slow down as our 'high' was steadily replaced by a kind of lethargy that descended upon us. Perhaps this occurred because the regular daily demands required on our voyage were not needed now in our more static life or perhaps it was just the effect of the heat.

However, two days later *Restless Wind* arrived and it was fun to see them again and swap stories about our first crossings of an ocean. Soon two or three other boats joined us in the bay. One was *Spellbound*, which *Restless Wind* knew as both were from the same Seattle marina. We had many BBQs on the beach and we were usually driven back to our boats in the evening by no-see-ums, little biters that were a real pest.

After some exploring we sailed with *Restless Wind* a few miles south to the small island of Tahuata. We collected our passports from the gendarme, told him where we were going and were relieved we would not need to deal with customs for a while. We were headed for Baie Vaitahu but on the way saw a rare yellow-sand beach, rather than a dark volcanic beach. We thought it was the Baie Hana-moe-noe. We anchored off for a while. Ashore we played games on the hot sands, swam and looked for shells. After an hour or so we realised we were getting bitten by no-see-ums that produced a nasty itch. When Randi got badly rolled and bruised in the increasing surf we suddenly appreciated the swell onshore had picked up and the rising waves were putting *Sky One Hundred* close to the surf line.

Getting off this beach and through the extra-short steep waves in our dinghy took a few tries but finally we got our act together although we were quite wet. When the four of us walked into the sea holding onto the dinghy there couldn't be any disagreement on timing the waves so that we could all leap into the dinghy, be on the back and not the front of the wave, and start rowing before the next wave swept us back on shore. Not all agreed with my timing and sometimes trips back to the boat made for lively, recriminatory discussions. On board we quickly up-anchored and moved to a safe anchorage in Baie Vaitahu for the night. Here we met up with *Merry Maiden* and a few other boats for four days.

This bay was surrounded by steep mountains and boats were frequently assaulted by huge blasts of wind off the slopes. Sun shades would be almost torn off boats or ripped, certainly if they were not attached by bungee-type cords. A good anchor system was required. In about 40' of grey clear water we had 120' of chain out with a 35 lb plough; we had been advised earlier to use all chain for anchoring and had joined up our two 60' lengths together with 150' of line. This would hopefully avoid any rope anchor line being cut by coral which we could experience. Jerry, from *Restless Wind*, swam over our anchor and checked it was embedded properly. In this windy environment we tied down every loose article, particularly the dinghy, with two lines.

Ashore there was a landing ramp, a small collection of houses, a school, two churches, some small food outlets, a football pitch and an unpaved footpath that led back into the woods. We felt very conspicuous walking around and wondered how the friendly locals thought about us. On an initial stroll through the houses a large woman beckoned me over and gave me some bananas while asking for cigarettes. While we talked she kept un-tucking her sarong from somewhere under her armpit, unwinding it and tucking it back again. I wondered just how far this unwinding might go! I wished I had worked harder at my French.

Jeremy and Erica with the children from *Restless Wind* met up with local children and had a great time boating with them and even had a soccer match. The locals played barefoot. Jeremy managed to score three goals but ached quite a bit due to his lack of physical exercise. A local man, Nickolas, wanted anyone who came ashore to try and fix his radio. I tried to help but later found he asked every yachtie. As visitors we were hard pressed to keep up with their generosity as we were given the odd fruit or vegetable and that they just accepted us in their small village. On Sunday they would come out to the yachts in their pirogues with their guitars and ukuleles and, if one was lucky, they would come aboard your boat and sing.

At this time a new French Governor was making visits to all the islands with his entourage. We were lucky enough to be here just as he was about to visit the village. Special festivities were being put on for him and there was great excitement among the locals. The girls were practising their singing, not that they needed to, palm leaves as decoration were being placed on the path from the landing ramp to the village, and foods were being prepared by various groups.

The Governor arrived in a French gunboat mooring close to us, and his entourage was ferried to the landing ramp in a small tender. It was quite a rough day with a good swell running and as they landed several of his dignitaries suffered a rather unfortunate "landing" being somewhat soaked by a wave which tried to race them ashore! The last Governor drowned in these islands.

All the yachties had been invited to the festivities which included food and dancing; an amazing custom. After various speeches and songs food was served including poisson cru and breadfruit, both out of this world. There were many faces looking in through the windows and doors and I was amused to see steady streams of full plates disappearing discreetly out through a back door. When the Governor was circulating Heather finally got her chance for a few words in French to him.

With the food finished, three fellows with two guitars and a bass consisting of a crude box, a wood neck with a string got going bashing out some fine rhythms. Locals invited visitors to dance but their rhythmic beat was so different from our western music that I'm sure one's local partner was happy for the dance to end so that they could couple up with a local. I was invited but loosening up one's body and getting into the rhythm of the faster movements took more than one dance. However, one could not but learn and be swept up by the friendliness and hospitality of these islanders. All the *vahines* (local girls) were dressed in their best finery and long dresses and seemed to sit in a group inside the schoolroom where the dancing took place while all the men (mainly young), dressed in their ordinary day clothes, fortified themselves with beer and vino outside – so what was new in this world?

Directly the music started the lads would charge in *en masse* and grab the girls for a vigorous dance; as the music finished the *vahine* would almost throw her partner's hand away and head back to her chair without any escort and the men would walk out. When the rhythm changed up a few revs to the heavy and fast pounding of the *'tamare'* they would rush in again, in more of a hurry now, and grab the girls who would be shaking everything every which way while the men tried to oblige. The often suggestive motions clearly indicated this island would remain habited. Once again just as the music got faster, louder and the gyrations quicker the girls walked off and the music stopped. The whole party was very impressive and most entertaining. The enthusiastic three-man band certainly produced a passionate beat. For those who did not wish to dance, there was an old 1950s French movie being projected on the side of a building. We conveyed our thanks to whoever we could and headed back to our boat.

After the fiesta we saw the Governor's entourage returning for their pick-up tender with some carrying their shoes and socks with trousers rolled up to very white knees. We shouldn't laugh as it was on this same landing that we got caught getting into our dinghy. Heather and I had just got in when a wave came in lifting the dinghy some 1-2 feet above the landing. On the way down the dinghy caught the edge of the landing which flipped it upside down throwing both of us in the water – along with our movie camera. As I viewed the surface from below I realised I could not get back in the dinghy quickly enough to save the camera, which was irrevocably damaged despite being in a plastic bag. I was mad with myself but one sometimes gets sloppy – and wet. The replacement cost in Tahiti was three times our purchase price.

While we were anchored there was very high surf and two

boats had their dinghies taken off the shore but luckily recovered. The morning we left we went ashore to say goodbye to some locals and tied our dinghy way up on the landing area to a steel rod. About half an hour later a friend came running to say our dinghy was adrift in the surf. It was recovered with the steel rod still attached to the painter. Loss of a dinghy would be bad news.

We had thought of sailing 40 miles south to Fatu Hiva, said to be one of the more spectacular of the Marquesan islands. We had also heard from yachties who had been there that it was an interesting place being pretty wild and unsophisticated, well at least in our western terms. It was also a source of tapu bark printings, a craft carried out by women. However, with our shortish schedule we decided against a visit. Instead, we left with *Restless Wind* to make an overnight 62 mile trip to Ua Pou (sometimes Ua Pu or Ua Pou but generally pronounced "why-poo") to the northwest. On the way we went into Baie Manamenu on the northwest side of Hiva Oa for the night but found it too rough. Our friends opted to push on overnight to Ua Pou while we returned to Tahuata.

Interestingly our Admiralty chart for these islands was printed in 1963 but parts of it were based on French charts from 1948. Judging by the engraved sketches which showed land profiles as one would see them, for identification, from the sea, I would suspect much of the data had been produced many years earlier. Reading some of the names of places was fascinating – Point Matakoo, Ua Huka, Ua Po, Eiao, Baie de Hakahetau, Opituha and Baie Hoatuatua were just a few tongue twisters.

We had a brisk sail passing through a series of rain squalls sighting first a small island, Motu Takahe, off the south end of Ua Pou. The crossing was uneventful while the island crags of Ua Pou, seen from 20 miles off, were very spectacular. The clouds, being continually on the move around the crags, provided us with new shapes every time we looked. We cruised along a very rocky coastline seeing an odd house or two and anchored in 50' at Baie Vaiehu where we met up with *Merry Maiden*, again. The bay was clearly indicated by a vertical vein of white rock on the east side of the entrance.

As it was not a calm bay and the wind was onshore Erica and I left Heather and Jeremy on anchor watch and rowed ashore. After spending 15 minutes finding a way through the rocky shore and surf – a hairy time – we walked into the nearest village where we saw the new French Governor again meeting up with the people. The grass skirted girls stepped forward and presented their leis and kisses to the main dignitaries. Speeches in Marquesan with translations were made followed by a beautiful and harmonious song of welcome. I didn't know how Erica felt but I felt quite emotional watching such a traditional scene. It was so different from anything I had experienced, being there in a small village with these delightful people going through a simple but dignified welcome. I felt a long way from home. There was the usual feast and dancing which we watched for a while. Before leaving one of the entourage told us *" we 'ave been doing this for a week now, it is very tiring and tomorrow we go to Fatu 'iva".*

While we walked back to our dinghy Erica expressed her view that it was very exciting but she would have liked to sample the food! Back on board we were met with groans *"Where have you been? What have you been doing?"* That night Heather and I took a bottle of wine and rowed over to the *Merry Maiden* for a chat. Seaton Gras the owner, from Boston, had five charterers with him, each for $400 a month. He was a talented fellow being able to speak the local language and also play the local music on his ukulele.

The next evening near dusk we were visited at anchor by two local French teachers who asked us if we could take them to another island. They came aboard and we asked them if they'd like a glass of red wine (we still had a good stock encased in flagons from our big shop in San Diego). We enjoyed chatting to them and when their glasses were empty we asked if they'd like another to which one accepted. Heather went below and poured him a glass (in near darkness) and came back into the cockpit; he took a sip and immediately spat it out! I quickly realized what had happened: earlier, I had changed the engine oil and placed the old oil in an empty wine flagon – yep, he'd taken a mouthful of engine oil. We never did take them to another island.

We were anxious to get our mail and so, next morning, we made a 30 mile beat crossing to Nuku Hiva anchoring in the Baie de Taiohae. There were about seven sailboats anchored out including *Restless Wind*, *Spellbound* and *Maredea*. We made new friends with Kay and Dave on their pretty ketch *Macushla*, the Larsons from Morro Bay in *Cherokee*, a single handed Aussie in *Lady Jane* and three jolly Aussies in an old wood boat called *Mink*. I had sketched their rather ancient looking boat one time when they were anchored close to us and it was pouring with rain. Our surprise when being invited on board was to find that half of it seemed to be filled with large vats of gin, rum and the like; we understood they would only speak Hungarian to French officials causing said officials, in frustration, to waive any necessary rules and regulations.

Later *Sundancer* came in. It was an interesting 60' steel boat from England with only a 10' beam, carrying a crew of young adventurers en route to Australia.

The wind could be off or onshore and there was always a bit of swell running. To combat that and to keep our boat at right angles to the swell and minimise any sideways rolling, we dropped our main anchor with 120' of chain in an offshore position and our 25 lb Danforth with 30' of chain aft on the beach side. At times over the next few days we would get some very heavy offshore gusts and I was impressed how this small anchor held firm.

With *Restless Wind* and other yachties a fine time was experienced with parties on the beach in the evenings. Getting back to our boats in the pitch darkness caused a lot of laughter trying to find them. With a swell running our dinghy would be lifting up and down at the side of our boat and the simple trick was to step off the dinghy onto the boat at the top of the passing wave, but generally this was a calm bay with good anchorage. The town had about 500 people whose houses stretched along the waterfront and into the valley behind where 3,000' mountains looked down.

The post office was just off the beach and each day we would enquire for mail. Receiving mail was a moment of great pleasure. Not receiving mail during times of sailing or while in isolated spots often led to significant build up of anticipation at any post office as we approached it. Erica always got excited when she received a hand-written letter but Jeremy might be a bit downcast if there was no letter from Linda. Picking up the post was really a big deal. Often the post master would only produce our mail after we had asked for it several times. We got into the habit of Heather trying first and then I would walk in an hour later. I never figured out why he was like that – was he concerned about catching fish in the evening or was he worried what his girlfriend was up to or was he just disorganised? After we left the island we had friends keep checking for our mail until one day he apparently exploded saying, *"I don't want to hear anymore about Sky One Hundred!"*

Once again our arrival coincided with that of the Governor. The festivities here were the largest, with a great display of native dancing, much hip wiggling in grass skirts and beautiful flowered head-dresses worn by both girls and boys. Some of the chanting and dancing was remarkably like that of

Maoris in New Zealand. A lady in the crowd handed us a paper cup each and poured us a drink — talk about fire-water! Wow! We had to go and dance our feet off. When the Governor left he came over for a chat and Heather felt very honoured indeed when he removed his gorgeous lei and placed it around her neck and gave her a kiss on both cheeks — it was then that we knew we should not have any visa problems.

This area had many local wood carvers. They were allowed to go with an inspector each year and select their wood, there being three types: sandlewood, tao and rosewood. The former was the most rare; a slight scrape of the surface would release the most exotic scent. In past days sandlewood lining was often used in dressers for clothes. The carvers were mostly located in a nearby valley. A hike there would be accompanied by hearing the gentle "tock" "tock" of carvers hammering a chisel on their latest creation. The carvers on this island worked on bowls, Tikis (their ancient Gods), spears and ukuleles. We bought a very well-carved tiki from carver, K. Pifrret, which we named "Gollum" and Erica bought a ukulele. It was interesting how their carvings were similar to those in other areas of the world. These hardwoods were very difficult to even drive in a nail and, apart from skill and artistry, would require considerable strength to work – unlike North American pines.

One Sunday after church, we were talking to a local couple, Simeon and Felicite Kimitete and were so excited to be invited to come for *kai-kai*, a meal. This was a great honour for us. When we arrived in their little area in the woods we were each presented with a beautiful exotic scented lei and then invited to eat *poisson cru*, fried plantains and fruits. The *poisson cru* was like no taste we had experienced. The white mixture of fish marinated in coconut milk and lime juice was quite unusual and a delight to the palate. We also had breadfruit, a vegetable which looked like green volleyball with bumps and tasted rather like a sweet potato. Our evening conversation was often interrupted by grunting pigs, bleating goats (one of which butted me in the calf while I was standing), tail-wagging dogs and purring cats and kittens. It was a delightful evening. Erica was very happy to be with the animals and was almost purring herself.

It was unfortunate our combined French was not enough for any serious conversation but the evening was very pleasant and relaxed except they ran their noisy generator for light. We had them carve an oval bowl for us; Simeon would carve and Felicite would sand the finish. Like many indigenous artists throughout the world, they did excellent work but never seem to move away from their traditional style. There was no western influence or change from their designs which I think could have perhaps made their wonderful carvings more interesting. Before leaving the island they came aboard *Sky One Hundred* and I gave them my power sander and sandpaper, and .22 bullets ("whirlies") which were valuable trading items for the yachtie. The bullets were used for hunting goats up in the mountains. They were very interested in Jeremy's carving of his 2' long ball and chain from one piece of Douglas fir, which he had been working on since California.

Jeremy and Erica and the kids off *Restless Wind*, wishing to leave their parents for a while, went off hiking in the mountains and stayed out for two nights. Jerry, Randi, Heather and I hiked up the mountain through a lush plantation of bananas but at the top we encountered wet weather just as the children came around the corner to make their descent. They had had a great time cooking up their meals and constructing an awning cover in case it rained. They had only met one person, a hunter, but all said it was great to get away from parents and boats! Erica felt a great sense of independence. We said hi and goodbye and they went on their way. We thought they were pleased to be coming back for their Mum's meals.

The weather being quite cool, we made ourselves a fire while we had lunch before returning. There was a great view of the bay and we saw three more sailboats had arrived.

Life was very relaxed in this bay. We were in the habit of rising at 6am and getting any chores done before the sun became too hot. We'd then go ashore for fresh bread and fruit. Locals were very generous and we were often given tomatoes, papayas, bananas, breadfruit, limes and mangoes. Staple goods were very expensive but, fortunately, we were still living on our stores. We had no eggs left and for some reason, while there were chickens everywhere, eggs could not be purchased.

On the way to the small adjacent village we encountered a number of 4-wheel vehicles and quite a few Vespa motorcycles. This area will develop quickly once an airport 60 kms away is built and a road through the mountains completed. The few shops supplied fruits, vegetables and canned goods. One owner had been found lying on the counter asleep; it was a pretty laid back place. The locals had a disturbing habit of dropping empty pop cans or wrappings on the ground and even on a shop floor. One had to be early at the bakery for bread as the supply seemed to depend on the number of yachties in the bay. On our last day we rowed in to stock up a bit because we did not expect to find supplies in the Tuamotus.

We spent many hours rowing across to other boats and sharing our various experiences and assisting anyone who had a problem. A number of production boats had problems. One had lost a mast and was looking for a replacement, another had found their forward stay connection was parting company with the hull. Others had leaks here and there so we were very pleased *Sky One Hundred* had no significant problems. Any problems we had were usually to do with some piece of mechanical equipment. Many evenings we would meet for potluck meals on the beach – once there were 40 yachties enjoying a get-together. Rowing back later in the warm blackness to our floating homes was always a delightful experience. We could then lie in the cockpit and gaze up at the brilliant heavens, satellites and blazing stars. The scent of Heather's Gauloise cigarette enhanced the strangeness of our exciting environment.

On another Sunday we were ashore early to attend the church service. It was a colourful scene as the locals were dressed in their Sunday best. We entered the church through a magnificently carved pair of doors. Listening to the harmonious singing was a delight. Even if someone went off key there would be smiles and chuckles. A row of girls standing in front of us were having fun as they sang, since one had high-heeled shoes. She would discreetly slip them off, nudge her companion who would try them on and then pass them on with her foot to the next girl.

Before we left this wonderful island we spent a day and a night at another bay on the south side of Nuku Hiva where only 15 people lived, including Daniel, a very good woodcarver, who was working on a very elaborate table when we arrived. His delightful wife, Antoinette, insisted on escorting us on a walk up a valley on an ancient path of flat stones which ran alongside a fast flowing stream. We stopped and watched some men chopping up coconuts ready for drying in smokehouses nearby. They would then be collected into sacks as copra to await the next schooner to pick it up for Tahiti. Another ten minute walk brought us into a clearing where there was a very small derelict church. Antoinette carefully opened the door for us to peek inside at the elaborate stained glass windows and everything that had been left in place including pictures and candles. We wondered when this had been built. Further on we stopped and sampled cool drinks of mountain water from the stream. Her dog, which had accompanied us, stood in the water up to his chin, taking large gulps of water every few minutes while grinning happily; he was so refreshed that he decided to chase a couple of pigs on the way back.

This bay was renowned for its no-no's, tiny black flies. We must have looked rather strange when we went to see the church as we wore long pants tucked into socks, with runners, long-sleeved shirts with collars turned up and hats – rather stifling gear but it kept the bugs out. These no-no's had taken us unawares especially near rivers or streams. The bites resulted in some uncomfortable nights of tearing arms and legs to pieces. We tried not to be too persistent as these bites, if scratched too much, had a habit of becoming infected; a few yachties ended up at the local hospital for a series of penicillin injections.

Tuamotus...........

"CRASH! CRASH! CRASH!
brought me and all of us out of sleep
in a big hurry"

After two weeks we said goodbye to our yachtie friends and headed southwest for a 490 mile sail to Manihi, with its many black pearl farms, and then to Ahe, two small atolls in the Tuamotus. They were both located at the north end of the Tuamotus, a range of many atolls, which stretched over 1,000 miles to the southeast, the largest archipelago in the world. Part of the southern area had been used for atomic bomb testing by the French, to which the Tahitians naturally had a strong aversion, and why not, with nearly 200 tests from 1966–1996!

One could circle around the north end of the Tuamotus en route for Papeete and avoid all of these atolls. However, we wanted to see some typical atolls in spite of the extra care and navigation required.

A key problem was that the highest part of an atoll was the top of a palm tree. This meant that at our low level in a sailboat they could not be seen until about 5–6 miles off; there were no clouds hanging around high peaks that could be seen from a greater distance. An atoll was mainly an area of near-surface coral generally in a ring shape. The ring would usually have only one useful pass into the inner area although there may be many small gaps. The whole ring was covered with individual groups of sand and palm trees and some underbrush called *motus*. Within the ring of *motus* there were open areas of water of varying depths interspersed with coral heads and small coral islands. The different shades of blue-green waters within were, without any doubt, unbelievably magnificent and spiritually uplifting.

On route one night, with a nearly full moon and Jeremy on watch, he called to me, "*There's a sailboat coming towards us.*" It passed about 100 yards away with no lights – like us. We called on the VHF and blew a horn but there was no response. It was kind of weird. Obviously there was no one on watch. Ships do pass in the night . . . .

Currents were said to be variable and sets were stronger closer to the islands. The northern portion our chart indicated "*westerly sets predominate under the influence of the S.E. Trades, with rates up to 1.5 knots*". Thus in a day one could be set forward 36 miles – enough to miss your island. I was taking star shots for greater accuracy as well as our usual sun shots. I was finding our dead reckoning was always less than where we were by celestial navigation. I managed a ham radio contact with Malcolm Wilkinson on *Meridian Passage* who was some 500 miles ahead of us. He advised they did not experience the current indicated by the chart; I was relieved to hear this news. We also heard on the VHF radio the boat *Spellbound* was behind us. This boat was later to experience some very tragic events.

We had very poor weather in the last couple of days just when we wanted good clear horizons to get sights and be able to see these low-lying atolls. We were experiencing extraordinary torrential rains during which the sea became a foaming whiteness; we had never seen rain like it. In the middle of the third day we spied some sort of low dark mass on the horizon. Then it disappeared while we had another squall for twenty minutes. As the rain stopped we finally sighted the tips of palm trees, leading to a cheer from the crew and relaxation by Heather and myself.

Our next problem was the entry to the only pass to get inside the atoll. Depending on winds and tides we had heard that sometimes the seas could make it rough to enter through a narrow channel perhaps no wider than 50 yards with coral on one side and a wharf on the other side. We had not been able to figure from tide tables what the tide would be but as we ran alongside the coral *motus*, the pass opened and it was quite smooth. We motored in through indescribably clear waters coloured with a multitude of different blues with greens. Were we arriving in paradise?

We anchored in calm water in the bay opposite what seemed to be a small hotel. About an hour later we noticed somebody swimming out to us. It was the pilot of the float plane that had landed earlier. Denis Roth, with Polynesian Air, hailed from Victoria. He wanted to know all about our trip, told us about his sail boat which he had built and invited us to meet him, his wife, Denise, and daughter when we arrived in Papeete.

Harold and Ursula on *Maradea* who arrived a couple of days after us were enjoying the atoll when one night we all had quite an experience. It was absolutely calm and the sunset was spectacular with an aura of varying yellow, red and orange colours. We went to our bunks with expectations for a good night's sleep in calm water.

CRASH! CRASH! CRASH! brought me and all of us out of a deep sleep in a big hurry. What was this sledge hammer-like noise in the bow? On deck there was a 20-25 knot wind and small, probably 1-2 foot waves were sweeping across the open expanse inside the atoll. Our boat was bouncing up and down in the waves; I figured our chain was caught directly below the hull in some coral and, being foreshortened, restricting the lift of the bow in the waves. With my flashlight I could see the anchor chain being smashed against the anchor roller. At the rate it was happening something would bust pretty soon and we would be swept onto the surrounding coral. Jeremy and I quickly got the dinghy over the side while I shouted to Heather and Erica to get out our spare anchor line.

In virtual darkness, I got into the dinghy and, hanging onto the side of the boat, pulled my way forward up to the bow while Jeremy held onto the painter; if we parted company I would be swept away. At the bow I took the spare rope anchor line and leant down as far as I could into the water and tied it to the chain. The line was then taken over the second bow roller and to the stern where it was tied down on the aft cleat. We then let the chain loose until the load was all on the rope and the crashing was stopped.

The idea was that a boat length of rope would provide sufficient stretch so that the rope did not get the same treatment as the chain. The trouble was that with each lift of the bow the rope would whizz for about a foot length over the roller and back again. Somehow I did not think the rope would last too long with this rapid movement. So, with the wind still gusting away, it was back into the dinghy where we repeated the whole process again and installed a second anchor line to the chain. Now we had the security of two ropes which, if they went, we still had the chain.

Next morning it was absolutely calm again. Looking down

through the clear water we saw our chain had, during the night, wrapped itself around a small coral head below us. We also learnt that the Schnetzlers in *Maradea* had had their motor on all night to reduce the stretch of their anchor line as they were nearly on the coral behind them. So this was paradise?

We stayed on exploring for a couple of days. Standing in one to two feet of water we could view with great delight the fantastic colours of the clams, whole spectrums of blue/green combinations, brown/orange combinations, all iridescent; little guys like Christmas trees one inch high that could disappear in a flash into the coral which itself ranged from blue and mauve to yellow and pink. There were fish with brilliant colours and sparkling turquoise water. It was as if we had our own private aquarium. A whole new world had opened up while school work was put aside once again.

Snorkelling was fun seeing shells of wonderful shapes and amazing colours. A bucket with a glass bottom would have been useful. While there seemed to be no sharks I always had an eerie feeling looking down and away into blue, misty, deeper waters. The super white beaches were more magnificent than those in Mexico.

The wind had swung from N and NE to E and SE and therefore the Ahe pass, which was on the NW side, should be without a big swell and waves. It was time to go to Ahe some 20 miles away. The entry experience was much the same as Manihi with another scene of brilliant and beautiful coloured clear waters. We dropped anchor close to the entrance and went ashore to explore. While going ashore in the dinghy there were 3- 6 foot nurse sharks circling around us. Jeremy was trying to hit them with an oar while I was yelling for him to sit down. They say they don't harm you but I did not have that confidence.

The following morning a power boat with some local fishermen came alongside and told us that the Chief of the atoll had invited us to a feast that evening and would come and collect us later. That this small atoll (23 x 12 kms) with some 560 people would invite all newcomers on boats to this feast was quite extraordinary. That we would be totally accepted in such a warm and friendly manner was a delightful concept.

We spent a few hours walking across the atoll to the outside of the reef and around the pass collecting shells. Later the same boat returned and dropped two young lads off to guide us to their village four miles away on the other side of the atoll. There being no random reefs inside the atoll as with Manihi and a nice wind blowing, we up-anchored and allowed them to sail our boat which pleased them immensely. They wanted liquor if we had any but were pleased enough to have some of Heather's cigarettes.

We dropped anchor close to *Joshua*, the boat of the renowned French sailor Bernard Moitessier. Later *Maradea* and Kay and Dave in *Macushla* arrived. After his anchoring we saw Dave standing waist deep on a coral head about 10' behind the stern of his boat; he would now have to re-anchor or experience a similar fate as we had in Manihi. Dave had been a refrigeration expert at Cape Canaveral; it was always a delight to be invited aboard and receive a drink that had an ice cube the size of half a beer can; something we had not experienced for a while and a real luxury for us.

We explored the beach and snorkelled in ultra clear waters. I also took a moment to row over to *Joshua* with a request. Our doctor friend, Duncan, had lent us "Heavy Weather Sailing" by Adlard Coles and I asked Bernard if he would sign the book for me before I returned it to the owner. This he kindly did. He and his wife had been living ashore on Ahe in a delightful village, growing their own fruit and vegetables for a couple of years. I got the feeling he was tiring of the environment which he did not find stimulating enough. He had urged the locals to try and improve their fishing techniques but with little success.

At 7pm we went ashore for the feast. It was to honour two French doctors who had come to Ahe to investigate the deaths of a number of babies. The apparent reason was that babies were often wrapped too tightly, perhaps following western ways, and died of heat stroke. It was puzzling to us why the cause had not been realised earlier by the locals; how many generations had this been going on for?

We were given heavily perfumed leis as we entered the small community hall and were seated at a long table occupied by the Chief, several male elders and the two doctors. The Chief made a speech of welcome and thanked the doctors for their help to which one of the doctors replied. We each had in front of us a fresh coconut with the top sliced off with a glass to drink the juice – we were amazed how much delicious juice there was in a coconut. The local ladies then brought in large platters of *poisson cru*, cooked meat and rice. This magnificent meal, much of which was cooked on or in the ground, was followed by large pitchers of steaming coffee. The local ladies did not join us at the table but sat around watching.

Soon the local band struck up with their guitars, drums and ukuleles with their powerful music and harmonious rhythms. It quickly seeped into and through the body and mind and, from that moment, perhaps the slight formality of the evening dissolved into a natural wildness. For a moment I tried to step back, absorb, and hopefully retain memories this enchanting evening in the warmth of the twilight, the caressing winds and the palm trees silhouetted in the setting sun; all in this wonderful exotic place with these friendly people. I realised it was very rare occasion, one that most people would never see and unlikely that I would experience again.

The dancing was thrilling with the rhythmic pounding

beat on the hollow logs. One of the young lads, who helped sail us over, had his eye on Heather all night. Some of the dancing by the locals, especially during the *"tamare"* where the girl dances almost between the man's legs, the actions of both partners left little room for one's imagination that their population would not decline. We poor westerners were all invited to dance, particularly Jeremy, but struggled to drop our simpler steps, loosen up and get with it!

After three days in this exotic place we decided to move on to Papeete some 300 miles to the southwest. Our route was to pass between two atolls, Rangiroa and Arutua, about 15 miles apart. The idea was to identify these atolls, sail between them and then, there being no further atolls, we would have a clear sail to Papeete.

However, as we departed through the Ahe pass the water was so colourful that I almost did not want to leave. I realised once more I might never experience this again, so I turned back and went back in through the pass again. This caused groans from Jeremy and Erica and comments the likes of *"Let's get going, Dad"*. There are times in one's daily life when there is a special and unusual high and this was one of them. I hung around for a while trying to embed the colours and atmosphere in my mind while still feeling rather loathe to leave; finally though, we passed through and out into the open sea accompanied by cheers from the kids.

The consequence of this delay was that as night fell we could only see one atoll on our starboard and had not seen the other atoll on our port. We ran close to the atoll to see if we could recognise any feature on the chart but saw none to identify it; close up all atolls tend to look the same. Night came swiftly upon us and I assumed, dangerously, that the starboard atoll had to be Rangiroa and I was too far away to see Arutua. So we started to sail on downwind to Papeete.

Without doubt, sailing *downwind* at night and not knowing exactly one's position, was a recipe for disaster. We would never hear the sea pounding on a coral reef downwind of us let alone know where to steer to try and avoid it. So very shortly, perhaps five minutes, when it was almost dark, red lights and alarm bells flashed in my mind. Apologising to the crew, I said we must turn the boat around and slowly tack back against the wind into open sea, the only safe place to be. This we did until light the next morning.

It was a most miserable and uncomfortable night sailing into the wind with thunderstorms, squalls and no moon but it was a safe sail. We had heard of some horror stories where sailors did not recognise a navigation light or a buoy and kept on sailing without double checking their position. In one case a large boat, motor sailing with a professional skipper, was swept onto a reef at night. Apparently, the current set shown in the Local Sailing Directions was not found as predicted, and the skipper's subsequent course adjustment made was not sufficient; a crew member later said, *"We might have heard the waves on the reef if the motor had not been on"*. There was another disaster near this area where a family in a 55' catamaran sailing downwind at night using GPS, who, after they had made a course correction to miss a coral reef, ran full blast into it totally wrecking their boat and had to be rescued. I could not imagine the utter shock of such an impact crash and knowing that I had *'blown it'* big time. It would surely be the subject of many nightmares thereafter. As we sailed downwind again in daylight and with a careful lookout we never saw either island so perhaps sailed between both atolls.

Papeete and Moorea............

We had plenty of wind over the next couple of days and Tahiti was sighted early and easily; it was good to have mountains to aim at again. Mount Orohena was the largest at 7,350'. The island was just as we imagined – very lush with exciting-looking peaks, steep valleys and with surf pounding on the reef.

We passed Point Venus where Captain Cook in the *Endeavour* had been sent from England in 1769 to observe the transit of the planet Venus across the sun. How did somebody in England know this was going to happen and who organised the trip early enough for Cook to get there? Accurate observations of this were expected to simplify the determination of longitude at sea. Heather thought that the measurements did not work too well as all the instruments were fogged when the crew saw all the beautiful Tahitian girls after being several months at sea. She also found it incredible that over 200 years ago a large sum of money was spent on the expedition and then not long afterwards, Queen Victoria did very little to help Queen Pomare of Tahiti when she was being given some "heavy" treatment by the French.

The entrance to the harbour was well marked and we soon found rows of sailboats moored stern-to at the waterfront quay. They looked jammed together side by side, intimidating enough to make us wonder if we could moor there too. After motoring around for a few minutes we decided it was time to push our way in. As we started backing in Heather let down our bow anchor along with a line attached to a floating plastic bottle to indicate where our anchor was amongst all the other anchors out there. We kept backing towards a small gap between two boats. Finally, their owners appeared and we squeezed in between a large trimaran from San Francisco and a large immaculately varnished ketch from Los Angeles. When close to the dock Erica and Jeremy threw lines to helpers on-shore who cleated them to bollards on the quay. A final adjustment of side fenders, a tightening of the bow anchor line and we were docked.

We sprayed the first foot of the lines ashore with DDT to hopefully keep cockroaches from invading the boat. We only ever had one onboard and Heather whapped him (it could have been a her) in a flash. We never brought cardboard boxes with food aboard as they often had the dratted things hiding inside. A landing plank was found for us so that we could conveniently walk ashore. It was always kept a few inches above the quay, again to prevent ingress of cockroaches.

It was interesting that, as we were docking, some of the helpers including Malcolm from *Meridian Passage* said, *"Welcome Patrick, we know you have got money someplace"*. We realised yachties must have been listening to Heather's mother asking on the ham net where we had further funds for expenses; she was running out of cash and wanted to know in which account it was available; a marvellous welcome indeed. Heather and I immediately had GATS to celebrate our arrival; in fact two as it was my birthday. We were then invited aboard other boats for drinks and to meet other yachties. There was no cooking that evening as we strolled, perhaps staggered, down to the *"truks"*, excellent mobile stalls on the quay which served fresh fish, chicken, pizza, steaks and super French fries.

We stayed this way for three weeks. It was an exciting and central place to be seeing the passing parade of traffic and pedestrians. First thing in the morning we dodged through the traffic straight across a dual-carriage road with shady palm trees down the middle to the post office to collect our welcome mail. Then the usual visit to the Port Captain's office where we found that, although we did not need a visa (we had entered on our British passports), we were still required to deposit US$600 each in a Papeete bank. This we had not anticipated and it seemed a bit pointless when we did not expect to be in the islands more than six weeks. However, the French Government wanted to make sure that if you were a pest or a liability they could ship you out *tout de suite*. Some yachties were fortunate to have paid their $600 in Canadian currency but had it returned when leaving in U.S. funds – the exchange difference was a good reason for a party. The Port Captain, Customs and Immigration were in one office and we found all their personnel to be very friendly.

Shortly after we arrived, a man from the FBI visited us. He wanted to know what knowledge we had of *Spellbound* and the crew. There was little we could advise except that we had been aboard once in the Marquesas and later had heard it was behind us coming to the Tuamotus. Apparently the parents of the family on board had been killed!

From our cockpit we had a grandstand view of the colourful passing life. Many people seeing our boat would say, *"I'm from Vancouver!"* and we would have a chat, perhaps giving them our mail to take back home. Apart from other yachties there was always somebody who wanted to come on board. Were they just curious or had they always had some buried dream to go sailing and see the world?

Once, the new French Governor was passing by and we welcomed him aboard. He was taken aback when Heather's pareo unwrapped and slipped to her hips as she gave him a helping hand across the narrow boarding plank – luckily she was wearing her bikini top. We chatted about his Marquesas experiences and he told us how exhausting it was.

Other visitors included Paddy Sherman of the *Vancouver Sun*, who gave us an update of events at home. Bill and Betty Slade from Ottawa kindly invited us back to their hotel for a swim, would drive us anywhere and would always bring a large sack of ice with each visit. They were on a six-month tour and told us about their exciting experiences on the Trans-Siberian railway.

During all the visiting, Jeremy and Erica would strive to keep up with their correspondence courses and deal with the latest comments from their far-away teachers. While Erica's teacher would paste happy faces and explain why she had something wrong, Jeremy's did not, leaving him, and me, wondering why he was given no such help. As a special bonus Jeremy and I could watch the beautiful girls passing by. They all looked stunning in their brightly coloured clothes and long black hair. Many flashed by on mopeds or bicycles with all kinds of riding styles. Seventy percent of the population was Polynesian and the rest were Europeans, Chinese and mixed groups which probably explained the outstanding looks of all the people.

The small town was quite old with many small, dark stores. Once we found our way around there were plenty of choices to aid in restocking our boat for the rest of our journey through the Society Islands to Bora Bora.

> *"One could just drop off the deck into the 80-degree water"*

Small buses were plentiful for travelling to any part of the island. Once when we were walking, a bus going the other way, stopped and turned around in order to take us the way we were going! At the town end of the quay there were "*les truks*"; past the other end of the quay there was only a normal beach and no wharf; here more sailboats were anchored out but with lines ashore. This area was called the "cheap seats" since crews had to use their dinghies to get ashore.

One morning, with a supreme effort, Heather and Jeremy visited the marketplace at 5:30am. They were particularly impressed to see locals buying live chickens and ducks, tying their feet and walking off. Heather was disappointed at the scarcity of fresh fruit and vegetables and the high prices.

On Erica's birthday, we were awoken early by the sound of gunfire. Looking out there was great excitement with the guns announcing the arrival of the steel four-master Chilean brigantine sailing vessel, the *Esmeralda*. She was 370' long and her beam the same as the length of our boat. She was the longest sailing vessel in the world. As she docked, the whole 300 training cadets, impressive in their "whites", were standing on all the cross arms making for a spectacular scene. Bands played and locals sang while the Tahitian girls were providing a wild display of dancing finishing with the usual exciting *'tamare'*. We went aboard and Heather invited two of their cadets aboard *Sky One Hundred*. We were later invited back for a personal tour and were given a bottle of Pisco and a copper ingot and I was able to try out some Spanish.

While the name *Esmeralda* is associated with historic scenes of sacrifice and courage, it was interesting to note this ship was apparently used as a jail and torture chamber during the military regime of Augusto Pinochet 1973–1980. I remember being in Santiago on a project shortly after the palace had been bombed; I was staying in the Hotel Crillon opposite and my hotel room had a line of bullet holes across the ceiling.

After *Esmeralda's* arrival, Erica had her birthday party in the cockpit with eight other yachtie children demolishing vast quantities of potato chips, cheese curls, ice-cream, pop and layered chocolate cake, junk food they had not had for months. On the subject of food or the lack of certain foods, I found that I had lost so much weight that my trousers were hanging on me like large bags. I went and bought a snug fitting pair of longs. Two months after getting back to Vancouver I found I could not even get my legs into them. So all you "over-weighties", sailing would be a very pleasant way to lose the pounds.

There was surely an ethic or custom with Polynesians that did not have a "yours and mine" or "your turn and my turn" philosophy. One got this feeling after a while. We were in a Tahitian bar for locals once when a bottle, and they were big, of Hinano beer arrived on our table. We did not know from where it came but apparently it was a welcoming custom to strangers. Locals often just gave you fruits or vegetables. One evening on another occasion in the same bar a large Tahitian came over and quietly suggested we leave the bar now as things might start to get rowdy – perhaps rough? He led us out the back door and we felt no offence. On so many occasions people just stopped to talk. Male bank tellers with a flower behind their ear was refreshing compared to our more staid way of life; behind the left ear meant one was 'taken' and behind the right ear meant one was 'available'. Perhaps behind both meant 'taken' but still 'available'.......?

In the evenings we would generally meet up with other yachties for a drink and chat aboard or go exploring leaving the kids to do their own thing, which usually meant meeting up with the *Restless Wind* kids; they had also come through the Tuamotus via the Rangiroa atoll. Sometimes Heather and I would play bridge with the couple on the adjacent boat. When we finished late in the evening and said good night we would leave the boat and wander down to the Pitate Club and join in with the dancing to the exciting Tahitian rhythms. The simple one-two, one-two steps of our usual dancing had to be changed to a "one-half step-two" rhythm which was much more lively. I had tried this at Nuka Hiva at the French Governor's welcome party and fumbled it occasionally but now at the Pitate after a few nights, I was enjoying this new rhythm. When a fast *tamare* rhythm came on we left that to those who could entertain us. Dancing with a Tahitian girl was to be exposed to quite a different motion, if not somewhat suggestive, but the usual very disconcerting action came from the girls who, just as the music was finishing, would throw your hand away as if to say *"That's it buddy"* and walk off. That's not to say one didn't have some out-of-line ideas but it also happened to the local guys.

These evenings were most entertaining. We were so thrilled and energised to be in this wonderful environment.

In the early morning, as we strolled back to the boat in the warm night air, we would be overwhelmed by the embracing aroma from plumeria (frangipani) trees and would collect their fallen petals. In our boat their scent was so strong and exotic – a wonderful catalyst which encouraged us, with a certain quickness, to retire to our forward bunk and continue our enjoyment. In the darkness we were quite conscious of *Sky One Hundred* straining forwards and backwards on her lines responding to the swell that came into the harbour. At night, too, with an ear against the hull we could hear a continuous crackling of what we thought was algae forming on the hull.

During our stay we explored much of the island, found beaches for Jeremy to surf on, visited museums, played some tennis and met many friendly people. We especially placed many plumeria petals in our head which was getting a bit high and needed a thorough cleanout with fresh water

We departed on March 23 and had a good sail across the 10 miles to Robinson Cove on Moorea, leaving the noise and bustle of Papeete behind. As we came through the pass in the reef we encountered utterly clear blue water. We could see our anchor and all its chain links 40' down. Actually we could see down at least 80– 90 feet probably because, there being little current, all sediment sank to the ocean floor. The water was so clear the fish and their shadows on the bottom could be seen in the moonlight.

We met up with Dick and Penny on *Active Light* again, friends whom we had not seen since November in San Diego. We spent a fine time with them catching up on their travels, playing cards and sharing meals. *Restless Wind* joined us a couple of days later. Another Fraser 42, *Idiot Wind* – not a romantic name – from Vancouver was tied up to a palm tree. We stayed in this paradise for two weeks snorkelling and absorbing the whole atmosphere. One could just drop off the deck into the 80F degree water, float around like a glider and look down at all the coloured fish and the occasional barracuda and shark. These normally awaited their prey in the pass although it was spooky seeing a barracuda or two looking at you. We tried some spinnaker flying from the top of the mast with Erica sitting in a rope seat at the bottom of the chute but the wind was barely enough to keep her from sinking into the water. The bay was spectacular with thin white beaches and lush forests behind, topped by massive irregular peaks above.

Erica said, *"We were fortunate to have Restless Wind to travel with. Their kids were good company for Jeremy and me and we also enjoyed Jerry and Randi and Dick and Penny on Active Light."*

We visited the Club Med there and were given free drinks but could not afford their meals. The One Chicken Inn had a dance each week and it was always my belief that afterwards it would be repaired for the next week's dance. We also visited another bay and anchored close to *Maradea* in front of the Kia Ora Hotel. We had a superb view across the reef to Tahiti. Heather, Erica and I hitch-hiked around the island in five lifts; it was a bit hot but we got to meet some local people and practise our French.

Perhaps the greater delight of travelling was the social contact over visual and other experiences. Clearly knowing some Spanish or French has helped. Knowing some Tahitian, allegedly easy to learn because of its limited alphabet and no tenses, would also help as some locals did not speak French or English. As with any travelling, without meeting locals and visiting their homes one can get an "outside-looking-in" feeling. The more intimate contact was something more readily available here especially if one had a lucky encounter or had some local person to visit. Staying more than one or two days at any one place was generally long enough to meet locals with mutual interest and they might even interrupt their normal activities to incorporate a trip, a meal or make a visit to your boat in that time; a stay of a week or more would be better

Imagine our surprise when we heard on the VHF a call from *Trondelog* to the Port Captain in Papeete. It was Bill and Edna Sibson and their two children, Tyler and Trish, who owned the garage in the middle of Edgemont Village where we lived! We did not know they were sailors let alone were on a trip to the South Seas. When they finished their call, we radioed them and said they must come to Moorea first (rather than the official visit to Papeete first) as we were leaving in a couple of days. So we upped the anchor, sailed out, saw them coming over the horizon, met up with much shouting and blowing of horns and escorted them back through the pass to our moorage spot at Opunohu, also known as Robinson's Cove and Papetoai. We had a great couple of days relaxing with our two families together.

Huahine.........

We had a short trip back to Papeete where we loaded up with food, water and fuel again, advised the Port Captain we were leaving and departed for the island of Huahine at 1645 hrs April 10. As Tahiti and Moorea slowly dropped from sight and night fell we all said a special goodbye to these two enchanting islands that were also the southernmost point of our trip. In quite lumpy seas the wind dropped completely and we spent the next nine hours motoring. After 130 miles we dropped anchor in six fathoms opposite the Bali Hai Hotel in the small town of Fare at 1300 hrs. We stayed here for five days along with *Restless Wind*.

We took a delightful stroll into town along the waterfront, visited the friendly *gendarme* and explored the very quaint town. Jeremy said it reminded him of a western town with its wooden buildings and raised sidewalks set back under the upper floors of the buildings. It was relaxing to sit under large, shady trees and watch the activity surrounding the weekly arrival of the copra boat. Sadly, at night when we strolled around, we saw through the windows the silhouette of heads in front of TV screens. We hoped this latest introduction of the western world would not signal the demise of the usual strumming of guitars. The windows were really hinged boards that came down over an opening and could be held up by a stick.

The next day we spotted *Sauvage* coming up inside the reef with Peter and Denise Roth aboard. He had swum out to us in Manihi. The following year they planned to sail to B.C. waters and settle on Protection Island, near Nanaimo.

We enjoyed the casual atmosphere at the Bali Hai Hotel where the locals seemed to congregate in the evenings. The girls behind the bar never measured liquor into glasses for drinks but generally just sloshed it in so quite soon there was a fairly happy atmosphere around. One evening we watched dancing displays followed by fire-dancing. Another exciting evening was followed by a spectacular display when the generator behind the hotel caught fire. Luckily a welcome deluge of rain helped douse the fire – it also helped fill our water tanks. Rowing back to our boat we could see hotel guests heading back to their rooms with lit candles.

One night Heather woke me up saying she thought she could hear a motor boat. There was a kind of drumming rumbling sound which I suddenly realised was not a boat but our chain AND anchor dragging across coral. I tore on deck and got the motor on while shouting to Jeremy to get up. There was only about a 15 knot wind but we were quickly drifting down onto a tall post which marked a coral reef. The bow just touched the marker post as I managed to reverse away; it was not very well attached to the coral as it just leaned over and disappeared! Jeremy pulled in all the chain while I kept us off the coral reef. Then we motored to where he could see a large coral clump below and away from the reef and re-anchored there for the night. Why our anchor broke loose was the question because the wind was not strong. Next morning as a local speed boat tore out from the local resort for the daily water skiing we watched as the driver stood up and kept looking all around for his marker post!

> *"It was more like an expedition into the Amazon jungle"*

Jeremy and I rowed out to the reef so that he could do some surfing. I became so intent watching him that I started to get too close to the waves. To avoid surfing myself I had to row "flat out", one of the advantages of the big oars we had on the Avon dinghy. The waters were really magical with their variety of fish and colours. Once when Erica and I were snorkelling along about 9' over a sand shelf, the undisturbed sand just in front of us erupted with an opaque swirl. I think our heartbeats kicked up a few notches as a 5' ray rose up with its big wings and lazily flapped away. At the edge of the shelf the sand sloped steeply and deeply into darkening and murky depths. A dropped shell would just roll down and out of sight. We both agreed it was eerie and would stay in shallow waters.

We spent the last day in Huahine about five miles down on the inside of the reef. Here Jeremy introduced us to a new activity, called the "Spin", to while away the day. He would take a halyard (a line to the top of the mast) to the stern, climb up on the pushpit and then, judging the roll of the boat, leap outwards and forwards with enough force to swing around the pulpit in the bow and land back on top of the pushpit on the other side. Quite a trick! Heather, always ready to try anything, managed to swing around the bow but was heading with some speed into the hull when we all screamed out, *"Let go, let go!"* which she fortunately did at the last moment, dropping into the water. (Heather said: *"Patrick was fearful I would damage the hull".)*

Raiatea and Tahaa……….

On our last night at anchor, a super brilliant sunset of rich yellows and orange silhouetted Raiatea and Tahaa with Bora Bora in the far distance. As we downed our late night drinks watching the changing of hues and colours it was calm and fascinating as the sun sunk lower. As we viewed the magnificent vista there was a certain sadness in a way, as we knew we were looking at our last two destination islands in the South Seas before we headed for northern climes.

In the morning, April 7, after buying some ice and fresh meat in Fare, clearing with the *gendarme*, we set sail for Raiatea and Tahaa about 20 miles away. These islands lay within a single reef. We had almost forgotten how to sail after spending time in these calm anchorages but day-hops such as this were very pleasant. We entered through the Passe Toahotu in the northeast and dropped the hook opposite Uturoa in Baie Apu on Raiatea. We went through the pass just ahead of the Abbott family in their trimaran, *Antigone,* and anchored in front of the local yacht club.

Leaving Erica on anchor watch, Heather, Jeremy and I rowed ashore to stretch our legs. We hiked up the inviting 968' peak, Mt Ohiri. The view from the top was superb. This was definitely paradise. Translated, Raiatea meant "faraway heaven" or "sky with soft light". We could easily believe this as we looked across the blue waters between the two islands to Tahaa which was topped in a varying spread of yellow to gold clouds stretching above us. I think we were all thrilled to be in such a fantastic place. It was also said to be the most sacred island in the South Pacific.

In the setting sun we could clearly see Bora Bora, 20 miles

away, oh, that mystical name! Uturoa looked like a toy town from aloft with its red roofs and a Tom-Thumb soccer match in progress. We were soon descending and came across a large grove of lime trees from which we collected a good supply of limes lying wasted on the ground – strange, as in the shops they were $1.40 for a bag of fifteen. We also collected some *guave*, which made a good drink when boiled up with honey and some lime added for a bit of "oomph".

Walking back to the boat, a car going in the other direction stopped and the driver asked where we were going. He was Jean Yves Guilloux, a school teacher, and his young brother, Cyril. He insisted we climb in and he would take us back to our boat. On shore we asked if he would like to have a look at *Sky One Hundred*. He declined as it was nearly dark, so we invited him for supper the following day. While keeping a look out for him on the shore the next day, we saw a sizable crowd gathering – he had brought his whole family! What an evening that was – they came with fish for *poisson cru*, fresh lettuce, tomatoes, a home-baked cake, drinks and music. I had to make a few ferry trips to get them all on board. This was an evening one does not forget. The simplicity of it all and the relaxed and fun outlook was a delight despite the fact that not all spoke English.

Next morning Jeremy and Erica led the *Restless Wind* crew up Mt. Oriri while Heather and I went off for some tennis. I also wrote a letter to Heather's dad, Chris, congratulating him on his retirement and concluded the letter with,

> *". . . your family here are all well but I think looking forward to being home again, it's nearly all northwards now! Jeremy is bigger than ever, getting cheeky and Erica is very pretty, logical and is damnably argumentative. Heather is a "karma" and continues to grow more beautiful, all the Tahitians dance with her. I think my hair is growing longer, below my neck now – it is certainly much blonder."*

Later, Jean Yves invited us to his house a number of times. He was 24, had his own two bedroom house, a lovely garden full of flowering trees, fruit trees and orchids, three cars and a colour TV. He had the latest *Time, Newsweek, Paris Match* and *Elle* magazines – what a luxury! At one dinner we had an icy rum punch then tender meat and French fries followed by *guave* pie and ice-cream. All this was topped off by a superb coffee. Sometimes I wondered how we could be losing weight. When we had recovered from this *repas* we were off to a local soccer club dance held on the nearby wharf – everyone in the town must have been there. There was an amplified band and once again we took to the floor and had a great time, strolling home around 1am, calling at the bakery for fresh, hot bread and devouring almost a loaf before we rowed back to the boat.

On another visit to his house his sister dropped in and she warmly welcomed us individually with hugs and kisses. When his mother and other members of the family were there they would sit around while we ate a meal. It was an incredibly relaxed and easy atmosphere. When we said we would be leaving soon we were persuaded to sail around Tahaa and return to Uturoa the following weekend when he would take us to the family plantation up in the hills to get yams, fruits and bananas.

So off we went and found some gorgeous anchorages off small motus where we swam and took the dinghy out to the reef. Jeremy found a beautiful textile cone shell but we were cautious, handling it with kitchen tongs, as we saw a small barb peep out every now and then. With the likelihood of this being quite venomous we dropped it overboard.

Then back to Uturoa where Jean Yves and Cyril picked us up in Jean's car and we drove up to the plantation on the west side of the island of Raiatea. It was more like an expedition into the Amazon jungle. Soon after leaving the car we were climbing through dense bush and the rain was beginning to get heavier. We nimbly jumped across a small stream, with Jean Yves leading the way with his machete. We beat our way upwards eventually reaching a grove of a fe'i banana (cooking variety) plantation. Erica was getting cold, so we both took shelter in a small hut while the others continued upwards passing coffee, orange and vanilla plantations. Finally, they found more bananas and yams and Jean Yves showed them how to collect the yams and how to bundle the bananas up Tahitian style.

When they came down, we all started down. Up to this time it had been raining for a while but was now thundering down so heavily that water was running down the bark of the trees. Worse, it was so cold that Erica, in just a summer top, was beginning to shiver and I had to bend over her to deflect the rain.

I was beginning to wonder about the small stream we had crossed on the way up and would have to cross again. My fears were confirmed as we found it to be a raging muddy, torrent almost thigh deep and getting deeper. It was noisy and quite frightening and not one I cared to cross. Jean Yves indicated *"no problem"* as he started to cross unsteadily with the banana bundle on his head. Half way across he decided to try and "javelin" his bundle to the other bank. No luck as it fell short and was whisked off downstream. He made it across along with Jeremy and Cyril who lost his thongs. I called out that we were going back upstream until we found a realistic place to cross. A couple of hundred yards away we found a spot with a fallen tree over the stream which I was prepared to cross with Heather and Erica. As we descended Heather felt she was trying to ski down the mountain on thongs. She soon abandoned her thongs and had lovely brown mud oozing through her toes as we waded back to the car in sucking mud.

Though Erica named this trip *"the fruitless journey"*, lower

French Polynesia

down we were able to collect juicy oranges, enormous avocados, guave, pamplemousse, vanilla and coffee. Pamplemousse which we had experienced all through these South Sea islands were like huge, delicious, succulent grapefruits. Back at the car it was hardly raining and Jean Yves explained that his valley had a micro climate which kept the valley lush and fruitful. I certainly agreed with that! Soon we were zipping along a coastal road and stopping to look at one of the many ancient *maraes*. These were made of enormous slabs of coral constructed like gravestones that enclosed areas about 20' by 100' full of coral layers, under which were said to be bodies originally sacrificed; very spooky as I'm sure we saw the odd bone sticking out – perhaps an effect set up by the local tourist board?

Although most of the islanders were Protestant, originally converted by members of the London Missionary Society who came in the early 1800s, they apparently still believed in the ancient spirits. This subject arose when Dick mentioned that they had entered through a pass in the south east of Raiatea and Jean Yves asked them what it was like as he believed it to be very dangerous. Apparently when the missionaries had entered through this pass the *maraes* on either side had sunk and, not so long ago, a canoe that had sailed from Hawaii was coming through the pass when a very strong wind had come up making it very difficult for them to enter. Jean Yves said he would never use that pass. Such was the birth of myths? The commune of Taputapuatea in the southeast was once the largest marae complex and religious centre in Eastern Polynesia.

At his house, we had luxurious hot showers, shampoos, and were given clean clothes while ours were washed and dried. We had lots of laughs about our trip while downing a couple of rum punches. Then followed a delicious dinner of roast chicken, more colour TV and a game of Monopoly with

French properties before we retired to our boat.

Next day, a Sunday morning, Jean Yves was at the boat at 4:50am to take Heather to the market where she purchased fresh lettuce, eating bananas and cucumbers. There was little produce as the farmers were still feeling the results of cyclone "Diana" which did so much damage, evidenced by roads and bridges washed away. Back at the boat Heather made pancakes for Jean Yves and Cyril but were told they only ate them for tea! Jeremy and I were quite happy with that. We then visited the eastern part of the island and saw more *maraes*. After a refreshing swim in a local river we all returned to the boat for dinner. Then it was off to the Bali Hai Hotel to watch some fire-walking by villagers from Apooiti followed by some dancing. The crew were not fully convinced about the fire-walking as the rocks did not seem particularly hot and felt a sprained ankle was a greater danger. We were not allowed to inspect the rocks.

In the middle of one night we were very annoyed to have a large Australian three-masted charter boat hit our boat with a thump. She had let go her shore line to anchor off for the night. In the darkness we saw no apparent damage and the skipper was very casual about it saying it could have been worse. He then went off to re-anchor. We were in some 16 fathoms, an unusual depth, at the time. Later, when I took a look in daylight I found black marks of his tire fenders all over our white hull. They would require a lot of effort to remove and I was more than annoyed. I rowed across to his boat and called for him and in front of his various passengers denounced his sailing abilities in fairly clear terms. Satisfied, I rowed back to *Sky One Hundred* while mentally changing his boat's name from *Golden Plover* to *Golden Plunder*. We later heard she blew two top sails just behind us on her way to Bora Bora.

It was time to be moving on again and we said our sad farewells to Jean Yves and his family with our thanks for their warm hospitality and said that we were departing with very fond memories of them all. We were given T-shirts, shell leis and a large stock of eating bananas. We kept in contact with Jean Yves for a few years afterwards. Back at the boat we could see the beckoning 2,400' mountains of Bora Bora just 20 miles away. It was very exciting to be close to the famed island, the source of many of our dreams about the South Seas. How had we ever managed to get here and what would it be like?

But first Jeremy and I had the task of heaving up by hand our 120' chain and a 35 lb anchor, nearly 150 lbs deadweight. It was a heavy task and I longed for a chain lever stopper which would allow us to stop and rest from time to time; not a good scene if there had been waves and we had had to move out quickly. At last it was up and we were on our way. We hit a bad squall just before leaving the pass which cut visibility down but we soon picked up the marker posts and went onto a good reach. Gradually the wind picked up and we realised we were in for a good blow. Jeremy and I quickly got a smaller genny up and after some effort put a couple of reefs in the mainsail.

Bora Bora . . . .

Bora Bora looked quite spectacular with the ancient volcanic peak of Mount Otemanu rising 2,400' above lush greenery. We could see the ominous and warning spray of waves on the massive reef which was about two to three miles off the south side – a very dangerous hazard. We were to later encounter a scary and dangerous experience on a reef here. Once aground on a reef a sailboat was virtually lost forever; there would be little chance of getting back to sea. We heard later that a sailboat before us had gone aground here and the locals came out to help the single-hander save as much equipment as they could. Understandably they were somewhat upset that he apparently never offered them anything in return.

It seemed a very fast trip and entering the Teavanui Pass on the west side, we were soon inside the reef and anchored in ten fathoms alongside *Restless Wind* again. The following morning we left Jeremy and Erica doing their school work and rowed ashore to meet Alex. He was the owner of the almost completed small yacht club to be formally opened May 5. We had an old West Vancouver Y.C. burgee and an old faded Canadian flag to give him for decorations. We walked quite a way into the town of Vaitape passing the Club Med. At the post office where the crew of *Restless Wind* thought they had seen three letters for Jeremy, we unfortunately found none. Poor Jeremy has been jinxed as far as mail is concerned. On to the *gendarmerie* to advise of our visit; the officer had three "pips" on his shoulder, presumably a captain, and was very serious compared to others we had met. We picked up some bread and headed back to the boat in the rain. We had, over the past weeks, experienced torrential downpours and the wind was howling about 40 knots from the north by the evening. We had earlier seen *Der Kormoran* leaving for Hawaii and this would not be good for them or for our next ocean voyage northwards – also to Hawaii.

We met up with Dick and Penny again, and did a 20-mile cycle ride around the island, stopping about half way for a picnic lunch and a cool swim on a white sand beach. At the Bora Bora Hotel we had drinks and a delightful couple from Seattle invited us to their over-the-water bungalow for more swimming. They showed us how to feed the fish with bread eaten out of our hands. Afterwards we had great hot showers before we left. Another stop, this time at the Club Med where we were given free drinks, completed a wonderful day.

One day, with Dick and Penny, we sailed and anchored near a reef in 10' of water. We wanted to go out and explore on the reef. The breakers on the outside were huge and impressive; we could hear them pounding down. It was an astounding sight and we wondered about the state of the seas. As we rowed our dinghies closer to the start of the coral reef we experienced currents against us and ended up getting out and pulling our dinghies. Finally, we tied them to coral and started walking on the reef. Water was flowing against us across the coral and it should have been a warning.

After a while our exploration of these colourful waters came to an abrupt halt when a huge breaker crashed down, the 49th wave, perhaps the 149th, and we could see a 1-2 foot high wave tearing towards us. We yelled, "*Get down, hang on tight to the coral.*" We all survived except for Penny who got tumbled resulting in some nasty coral cuts and scratches and the loss of a pretty ring from her finger. End of exploration and back to our boats to repair the wounds and have a couple of GATS.

"Every day in port is one of relaxation when we are not on the move. One does not hurry anywhere and sometimes we move so slowly that, in a large city, it might almost be considered loitering. Morning tea starts the day followed by a quick look outside to see whether boats have arrived or departed. Breakfast of pamplemousse, papaya and pancakes or an early visit to the market before it gets too hot will be next. Afterwards a couple of hours of school work (amid groans) while Heather and I may shop, write letters, change engine oil, wash or clean up. Social visits, "rapping" and planning excursions with others happen at the drop of a hat.

Usually there will be an early morning swim and maybe a hunt for shells (tracks were often seen in the sand). We may move to another position but only generally between 10am and 2pm when it is easier to see any coral heads. Creative urges surface from time to time with Heather making Ha-

waiian shorts, a dress or swim suits, sketching, rope splicing or improving systems on the boat. Midday it is pretty hot and is often reading and "rapping" time. In towns there seems to be a considerable urge to visit and spend money, something we can't do offshore, Thank Heavens! Most activity definitely seems to revolve within a half mile radius of our boat. Visits to points of interest way off are not too frequent as they require a lot of effort to move in the heat. Sometimes the kids will go on one boat while all the adults will congregate on another.

As we watch the sun go down, we will decide whether to have supper aboard, stroll out to some cheap spot to eat or maybe have a BBQ on the beach. It is a pretty lazy life in all, which some people seem to be able to adopt indefinitely; I'm not sure that I could. I'd miss our home friends, the ready availability of ice, running hot and cold water and armchairs".

One day we met up with Sue Fleishman, a writer for *Bon Appétit* magazine. She was interested in doing an article titled "Recipes and Shortcuts from Seasoned Sailors" and within a short time was in deep discussion with Heather on her experiences providing food for us all. While that was going on, I took her husband out to our boat to see the set-up we had in the galley and elsewhere. Here is the essence of the published article:

"The trip was a learning experience for the first mate and the captain," says Heather. "The galley is well centered for stability and has an excellent propane-powered stove, but if we do a trip again like this we'll add a generator for good refrigeration. Our five cubic foot ice-box with its airtight lid and four inches of insulation held the cold for only ten days; after that we sorely missed it. Two five-gallon containers of frozen water first served as a refrigerant and then served as *a supplement for additional drinking water. Covering our ice in towels helped to prolong its effectiveness.*

Since rough seas slowed us and we had no luck fishing, I'd fortunately planned menus and stocked up on canned and dried foods for more time than we anticipated being at sea. As a precaution we marked our cans with wax pencil before stowing them in the bilge. Dry staples wrapped in foil kept well in high lockers. Our eggs, coated with petroleum jelly and packed in hard plastic containers, stayed fresh for six weeks. I would turn them occasionally so the yolks would not stick on the sides."

The Hills festooned their cabin with hanging open slings for oranges, fruit and potatoes, which kept fresh for at least 20 days. Hard cabbages and green tomatoes were other good travellers, perking up canned vegetable salads. Bananas unfortunately would all ripen at once, so Heather invented breads and cookies to use them up. She found her pressure cooker to be a great bread oven.

Sue then added Heather's marvellous bread recipe which whenever produced at sea was almost always devoured instantly by the crew:

HEATHER HILL'S SALTWATER PRESSURE COOKER BREAD – MAKES ONE LOAF

1 envelope dry yeast
1½ cups sea water or salted water (105 to 115 degrees F.)
4 cups all purpose flour
1 tablespoon sugar
Flour

Lightly grease and flour pressure cooker. Dissolve yeast in ½ cup sea water in large mixing bowl and let stand for 10 minutes to proof. Add remaining water, flour and sugar and stir until dough is consistency of heavy paste. Lightly flour hands. Turn dough onto lightly floured board and knead until smooth and elastic, about 5 minutes. Transfer to pressure cooker, turning to coat all surfaces, and cover with lid. Let stand in draft-free area for 2 hours. Remove petcock (steam valve) and place cooker over medium heat for 15 mins. Reduce heat to low and cook an additional 15 minutes. Turn loaf over, cover and continue cooking over low heat for 30 minutes. Cool before slicing.

After nearly two weeks on this fabulous island we were reluctant to leave but it was time to move on. It was northwards but first a visit to the *gendarmerie*. Here we collected our deposits made earlier – for good behaviour? and advised our imminent departure. One thing about Customs and Immigration regulations: they will always differ from port to port in any country. We always read that clearances were needed so that they could be seen in the next country but nobody so far had looked at our U.S. or Mexican certificates and I wondered where that little green card was that I signed when we left Canada.

Our last evening on Bora Bora was memorable as it was the opening of Alex and Michelle's Bora Bora Yacht Club. The evening started with a splash. Being built out over the water with no handrail yet, one cruiser, when shifting his chair, got one leg off the deck and, much to the amusement of us all, disappeared over the side into the sea while still holding his drink. Many cruisers and locals attended the party and it was a wonderful evening with much music, some of the best Tahitian dancing we had seen, drinking and singing. This was what it's all about!

We shall never forget walking back to our dinghy on the shore nearby where we were serenaded by three young boys with the most wonderful voices. It was very moving and also very sad to realize that we were now leaving these beautiful islands behind us. As we slowly rowed back to *Sky One Hundred*, a warm offshore scented breeze caressed us while we were both lost in our individual thoughts and our time with these friendly welcoming Tahitians. The water flashed from the slow stroke of our oars as the exotic rhythm and singing of the locals followed us. Our time in this seductively embracing climate with islanders was at an end but we would have our memories forever and what more could one ask for? Could there be something closer to paradise? I doubted it. We had the good fortune to experience a new perspective and learn something of another life, the way of living on these South Sea Islands. The following day we were to commence our return trip and back to our way of living; this we would do with a better understanding of people and life.

RAIATEA
"sky with soft light"

Heather's 'taken'

Bora Bora in sight

Visit marae with
Jean Yves

French Polynesia

Ugh!

49th wave

Dick, Penny and us

Farewell to Bora Bora

Hanalei Bay sketch

Champagne from Jerry and JJ
Farewell from *Restless Wind*

French Polynesia

Heather
calculating
our position

Alaska in
sight

Back in calm
waters

Home on the Waves

Quadra weather ship

La Perouse glacier

Reid glacier
Note person

Lamplugh glacier

French Polynesia

Calving ice

Seals on bergy bits

Baranhof hot springs

Killer whales

Mamalillicula

French Polynesia

DONE IT!

193

CHAPTER 9

Rough Passage to Hawaii

"Suddenly high above us houses perched on the hillside peeked through the mist"

On May 7/78 with heavy hearts and fond farewells to cruising friends, particularly Dick and Penny aboard *Active Light* who were continuing on around the world and the Jacobs family on *Restless Wind*, we hauled in our anchor and made our way out through the Teavanui Pass ready for the next passage – to Hawaii. We always felt a bit uncomfortable "going to sea" again especially since for the past couple of months we had only been doing short passages between islands. Now we would be on the move for at least three weeks to travel some 2,400 nautical miles (4,800 kms or about Vancouver to Miami). It could take longer but Jeremy and I had scrubbed the bottom of the boat removing all barnacles and sea growth off the hull. I did the top half from the waterline down and

Jeremy earned his keep doing the bottom half – the hard bit! To not do this, significant extra growth would gather en route resulting in perhaps an extra 4-5 days travel time. In fact one boat took over thirty days after developing considerable growth on its hull. We had experienced hard stalactite growths some 2-3 inches long on the stern overhang.

As we cleared the pass, my thundering headache told me I had had far too much to drink the night before at the new yacht club. Heather and the kids were fine but sorry to leave. Heather had stocked the galley well. In particular we had found an isolated lime tree with hundreds of limes lying on the ground. These we collected and later spent a day squeezing them into about ten bottles. An aspirin was dropped in each bottle and the lids screwed down. The aspirins acted as a preservative and at the end of this trip we were still drinking fresh lime without any sour taste normally experienced after a couple of days of them being cut open.

We were heading north to Hilo on the Big Island of Hawaii. The winds were from the northeast but we eased off into a close to broad reach in anticipation of picking up a southeast wind. In the days of the square rigged ships the target when travelling north was to gain as much easting as possible in the southeast trades until about half way to Hawaii, meet up with the northeast trades and then ease west again.

Unfortunately this was not to be for us. We did not experience any significant southeast winds, only northeast winds. We later found other boats had the same experience. For six days we had strong northeast winds and no southeast trades as the Pilot predicted. We got a quick sighting of Caroline Island four days out which gave us a confirmation of our position, always gratifying. We all gritted our teeth as we slammed our way north, gradually making some easting to 148 degrees to make our trip from the equator to Hilo more comfortable. The only one who really enjoyed the trip was Goldfinger who was in his element keeping us right on course.

It was a rough trip. Here is a description of my off-watch-in-bunk experience written on watch part way through what turned out to be an 18-day trip:

OFF WATCH

"My body lifts, hangs suspended then drops, compressing foam. With any position sharp sideways thrusts are imparted at any time causing limbs to roll and shake. Water gurgles, hisses and surges past my ear twelve inches away – one inch of hull between me and the ocean bottom two miles down. Body is lifting again, foam compresses, wave passing under, boat drops with a whoosh and a final twist. Galley equipment rattles, a halyard resonates with irritating intensity, the wind whirs. Boat hits smooth patch, accelerates then slows suddenly running into an oncoming wave, body responds with a longitudinal jolt. Boat lifting again, foam compressing, then a hanging feeling followed by a whamming bang as the boat drops over a passing wave. Boat slows and then accelerates again.

I wonder where did the southeast trades go? We have been beating to weather now for six days, Bora-Bora to Hawaii. A big bang right at my side startles me and spray crashes on the deck and cockpit. Jeremy on watch curses. Boat lurches up and over a short steep wave with a sharp snapping motion. Boat accelerating, lifting, hanging and crashing down with a bang which makes me think of "oil canning" – the inward flexing of hulls! Another wave strikes and a heavy cascade of water falls on the deck. See it streaming across the leeward ports. Watch odd drip fall off end of forward hatch bolt; never seen that before, or this droplet moving slowly across ceiling – now where's that come from? Body stiffens from sudden twisting lurch, brace body against lee-

board. Wind whirring increasing. Hear watch adjust helm slightly to move boat more off the wind; hope we'll recover easting when it gets a bit calmer.

We are doing about 5-6 knots with a triple reefed main and a 90 square foot storm jib. Do not envisage basic change in wind direction therefore this should be the longest starboard tack for us — some 2,400 miles. Realize I am holding head off my pillow so relax body again. Sharp, snapping, pitching motion, then acceleration, general humming noise amid all other racket. Boat lifting, whooshing down with a twist. Suddenly no sound, no movement, nothing except the ticking of the log. A tenth of a second later — wind wailing, water hissing and banging and we are back in the groove again....... boat lifting, foam compressing, suspension, whoosh....... accelerating, boat lifting, foam ..zz... compressing ...zzzzzzz..... whoosh .... zzzzzzzzzzzz ..... Suddenly, a series of violent shakes on my shoulder and a grinning face above is inviting me to partake of the whole scene from the cockpit for the next two hours!"

My notes remind me:

I was quite sick after the first day and queasy for the next five days. Our skin started to get quite oily and greasy. Sores and bruises appeared on our elbows and bums due to continual motion, sometimes rough, from northeast swells, and the loss of body fat. Some of us were getting constipated and use of an enema was considered. While Erica was suffering the most, she found this the easiest passage as there was none of the rolling that was associated with the downwind sailing. Heather was performing magnificently and held the ship and crew together. Our bottles of crushed limes were providing excellent drinks.

Sky One Hundred was performing well under these rough conditions and was maintaining over 5 knots which was our intent. It would sometimes sail for two days on its own except when we moved off the wind to have less movement for supper, sail changes or even using the head. On the foredeck there was a fishy smell in the sails. In any calmer moments we would wash on deck. There was a noticeable deterioration of varnish. Coming on watch at night our bearing could always be checked by the position of the Southern Cross against the backstay. Our ham radio was consistently used for weather conditions, to chat to other boats and to patch through to Heather's parents through the use of the DDD net in Vancouver. Much of the time we kept in the forward end of the cockpit under the dodger to keep out of minor spray.

Life on board was not the easiest, consisting mainly of eating, sleeping, reading, happy hours, navigation, sail changes, checking equipment and running the engine for short periods to keep the batteries fully charged. While Jeremy and Erica tried to do some of their school work it was a difficult task for them. The happy hour was always good for a chat on the food we would purchase on arrival in Hilo – ice-cream being at the top of the list, job allocations (not so popular with Jeremy and Erica), discussions on reading material, homework and, often, how many days until our next landfall! After a while on this uncomfortable passage we all agreed passages were definitely for the birds; we had yet to meet a boat that enjoyed passages.

We crossed the equator this time with less ceremony, but with an extra drink at "happy hour" as we listened to a Montreal/Boston hockey game. Once again we found the temperature cooler a few degrees north and south of the equator. The weather was bright and sunny but there were no doldrums.

We saw very little marine life on this leg, only a few porpoises, frigate birds, boobies, terns and schools of fish. Sometimes we put out a line with a 4' steel leader to prevent the heavy duty line being bitten through but generally the process of gutting and cleaning up was too much effort. Jeremy was "thwapped" by a flying fish as he sat in the cockpit one night – he actually thought I had thrown a dishcloth at him. Fish we found on the deck were too small for the frying pan. At times we have seen a school of flying fish in flight above the waves for 30-50 yards – very impressive. Sir Francis Drake wrote *"of the length and bigness of a reasonable Pilchard having two fins reaching from the pitch of the shoulder to the tip of the tail . . . whereof she flyeth as any feathered fowl in the air"* and thought they could fly for at least a quarter of a mile at a time.

There was a big load on Heather who was frequently the referee between Jeremy's and Erica's wants and demands. Sometimes I could see the pressure building and would have to move in and take some of the load or just direct the kids to ease off and fend for themselves. We have actually heard of a wife who had to lock herself in the head for some peace, isolation and relief from the close quarters of her family living on top of each other. Generally though we all recognized that there were unwritten periods when you did not disturb others because it was their private time and, for the most part, life proceeded on a nearly even basis. This was not one of those passages where one could retire to one's bunk or even take a blanket and find solitude on the forward deck.

We saw no ships on this passage, just blue seas. There was the usual daily check of the weather, our sun shots and calculation of position, and a check of the boat lines and sails. There were ham radio calls to Vancouver and other boats. All this did not leave much time except for a short read, some writing, eating and then night was closing in. One day, when we attempted to start the motor to charge the batteries, it simply would not start – a major problem. I figured it must be the starter solenoid again. It required removing and unsoldering the connections to open it up so that all the carbon could be removed from the connection plates. Just imagine un-soldering and then re-soldering again while the boat was heaving around! Without our Honda generator this could not have been repaired – better to have had a spare starter solenoid or two. It was not a pleasant task with all the boat movement.

Finally, we were closing in on Hilo. Each night now we had the boat pointing more up into the wind to ensure we kept to the east of Hilo and did not get downwind of the Big Island. At this time celestial navigation was getting less possible as we were experiencing considerable sea mist, and while we could just see a misty sun at times, we could not see the horizon to obtain the sun-horizon angle. Fortunately, our Zenith radio with its direction finder was picking up the Hilo radio station. As the beam moved perpendicular to our port bow we steadily turned and headed in that direction. The highest peak, Mauna Loa, on the Big Island of Hawaii, was 14,000' and we could not see it. As we edged through the sea mist we kept looking for the shore but could see nothing but thick mist. Suddenly, high above us houses perched on the hillside peeked through the mist and then right in front of us was the entrance to the Hilo Harbour. What a relief after 18 days' sailing!

As we entered the harbour we saw *Der Kormoran* from Victoria leaving, and across the harbour the Maple Leaf flag of *Oriole*, the Canadian cadet training ship which was taking part in the Captain James Cook bicentennial celebrations.

At the dock we were back with bureaucratic reality as a friendly customs official, Jack Cooper, came aboard. We were advised we could stay at the dock for three days free, but we had to dispose of any fruits and vegetables including our taro, eggs and prescription drugs. We could keep our 28 remaining eggs so long as we boiled them in front of him. This we did but did not wish to see a boiled egg again for quite a while.

Our great delight was going ashore, entering an air-conditioned supermarket, and picking out all the goodies we had craved for the last three weeks, the first of which were ice-creams. For three days we enjoyed being on land again, having a game of tennis, relaxing, visiting black beaches, hiring a car and making a visit to the volcano Kilauea. At the latter it was very impressive to see the extent of lava flows and damage that had occurred.

When leaving on the third day for Lahaina on Maui, I found our reverse gear would not work and had to return to the berth. An inspection of the hydraulic transmission indicated the plates were not engaging – probably as a result of sea water getting into the transmission way back in Mexico. So Jeremy and I had to take the transmission box out while Heather re-hired a car. We drove 100 miles to the Kona coast where there was, fortunately, an outlet for Isuzu diesel engines. Two days later we had the box back in place but then had to pay for the three free days plus the extra days. It was an expensive and frustrating time but it did allow us to see the sun rise as we drove through the cane fields and the cattle country of the huge Parker Ranch. While waiting for the repair work we drove down to Cook's Bay where history informed us Capt. Cook was murdered on the beach 200 years ago. There we bumped into the CBC group photographing the *Oriole* at anchor, and met Neil Sutherland, a superb pianist we had met years ago at a party in Vancouver.

After looking at the rolly anchorages on the east coast we decided to sail straight to Lahaina. Following payment of our heavy wharf fees we had a firm overnight sail around the north end of the Big Island, across Aleneuihaha Channel and along the south side of Maui.

Strong winds were experienced as we passed the mountain gap at Kihei. As we arrived at Lahaina there was a most fantastic display by a couple of hump-back whales, 200 yards away, which leaped completely out of the water and made some graceful dives.

We were horrified to see so many boats anchored in the channel outside the reef. This did not bode well for finding an inside berth away from choppy waters, and meant we would need to dinghy ashore. Knowing the harbour had limited berth spots we entered on chance and managed to find a temporary place to berth. The friendly Harbour Master helped us tie up, while we explained where we had come from and invited him to lunch at the Lahaina Yacht Club. He said that night we could tie up on the gas dock and luckily, the next day, he managed to find us a berth. This was a relief as the gas dock had been very noisy at night and the new berth allowed any of us to go ashore at any time, have showers, do more shopping and especially meet other boaters. Our buddy boat, *Restless Wind*, caught up with us here and we stayed for a week.

We crossed over to Lanai, the pineapple island, and moored stern-to on a wharf. We spent two days there relaxing on the white sandy beach, surfing, eating pineapples and enjoying BBQs with *Der Kormoran* and *Restless Wind*. We also watched about 200 very white-skinned boys having fun for a couple of hours, between picking pineapples. We had a great sail past the marvellous cliffs of Molokai, crossing the Kaiwi Channel and pulling into the Ala Wai Yacht Club on the island of Oahu. Our stay here at this most friendly and helpful club was incredibly relaxing. We were tied up literally under a palm tree. When the wind blew one could hear the palm fronds rattling together. It was such a pleasure to feel so good in this particular environment.

The upper balcony of the club overlooked the harbour and we spent many pleasant hours watching all the activities. Jeremy went off sailing in a race around the islands while Erica joined up with the *Restless Wind* children who had caught up again.

It was time to work on our boat which was looking a bit worn, so I re-painted the upper deck and sanded and re-varnished the teak toe-rail. Now she looked in good shape again. One of the tasks was to install what I called our rollbar. It was a one-inch diameter galvanized steel tube that I bought, bent round and installed at the rear end of the cockpit. Its purpose was to hold up a cover over the windward side of the cockpit as we travelled north to Alaska. The wind would be a westerly and colder further north. As we spend a lot of time in the cockpit we wanted some comfort from spray and wind chill. Later, this did a good job allowing us to stay and work in the cockpit and not have a closed-in feeling.

Heather spent a very frustrating day trying to obtain some Alaskan charts but none were stocked here. She ended up on *Greybeard* (here for the Tall Ships Race to Victoria) and the owner Lol Killam kindly lent her an Alaskan Cruising Guide. Later we met up with the Courage family from Alaska on their 37 CT, *Encourager,* who we had heard on the radio. They also lent us some Alaskan charts.

Jeremy and Erica managed to get some school work done and I admired their ability to put their heads down to work in this distracting environment. Jeremy also managed to get on a weekend race to Molokai just after we arrived. He sailed on *Bottomline*, a Morgan 42. He told us the stanchions were underwater on the beat across and took on about 500 gallons of water – pretty scary! A Farr 1 Ton decided to sail back and made the 40 miles in three hours. Erica earned some pocket money on a large boat heading for the Marquesas, by naming food can contents with a felt pen after peeling off the paper labels. Heather and I made a bus trip around the north side of the island and also visited the museum.

One evening when we were eating on the club balcony we saw a sailboat coming in through the pass. There was an unusually high surf running on the south side of the islands as a result of extremely strong winds and a storm to the south. The wind and surf reached peak conditions this particular evening and quite a few people were looking at the buoyed entrance to the Ala Wai Harbour. We could see there were many surfers and then this ketch was coming, virtually surfing in, with full sails, on three of perhaps the larger waves. Suddenly we recognized it as *Moon Island* with Colin Hempsall, a fellow club member, aboard whom we had last seen in San Francisco. He was returning after his single-handed trip to Samoa. We rushed down to meet him and help him into a berth. It was good both to see him again and that he had come safely through the pass – a few drinks flowed that night. Apparently the Coast Guard in Pearl Harbour had offered him one of their people to help him bring the boat over. It wasn't until they were actually roaring into the pass with this man at the helm that Colin realized he had never sailed a boat before!

Next day Colin came over and Heather made a "special" breakfast full of eggs, bacon and hash browns to fill him up a bit. We had both lost a lot of weight in our travels. Afterwards he and I stood together for a photo to show how much weight we had lost by each having one arm inside our shorts – our own shorts, mind you! Obviously sailing was an excellent way to lose weight and keep fit.

After seven days we reluctantly left this wonderful environment and headed up the southwest coast of Oahu to its most westerly headland. Here we waited in a small bay close to a cement plant. At 0100 hrs in the morning we upanchored and headed out in the darkness across the Kauai Channel so that we would reach Kauai in daylight. It was about a 75-mile crossing.

Disconcertingly, as we came out of the shelter of the island we ran into the full blast of the northeast trades and fairly large seas. Goldfinger was doing his job well but in the darkness we could just glimpse the shimmer of large waves and could certainly feel them. We were going well for a while when, on Jeremy's watch, there was a huge crash and Heather erupted out of the cabin couch and landed on me sleeping on the floor. Donning my harness I shot into the cockpit to hear Jeremy saying, "Holy Smoke, I was standing on the BACK of the cockpit seat for a moment!" meaning that *Sky One Hundred* had been laid over on its side and then righted itself. This was about our worst moment at sea.

We quickly disconnected Goldfinger and, for the next couple of hours, I steered keeping, as much as I could, my eye on the faint gleam of waves coming up high behind us from our starboard quarter. I suspect that the trade winds and the resulting waves were constricted by the nearby land and hence accentuated. We all breathed a sigh of relief as the conditions eased. In the morning light we could see land and shortly were cruising along the north coast of Kauai. After three or four hours we turned into Hanalei Bay, close to Club Med, and tucked in close to shore to minimise the effect of swell running in the bay.

This was an excellent place to finish our warm water cruising. There was a long yellow beach, with no surf and the whole bay ringed with palm trees backed by rich green covered mountains. *Restless Wind* arrived and joined the many boats waiting for their next ocean sail, mainly back to home in Seattle or Vancouver. I called it 'Banker's Bay' as, for many, money was running out, some had loans on their boat, and it was time to return to work. There were many parties on the beach and so many stories of adventures were told. Jeremy and I went into the water for over a day to clean the hull for our twenty-day trip north. At 70F degrees, the water was chilly after our normal 80F degrees further south. I did not like the murky water either. I always felt I would never see a shark coming although I do not know what I would have done about it if I did.

On the last night while the kids all had a party on one of the boats we adults went to the Tahiti Nui Club where we were able to hear the most harmonious and wonderful singing by locals, something we never hear on our western entertainment programs. We walked home in the warm night air with the brilliance of stars lighting our way. On the beach we said our farewells to our friends with hugs and kisses with promises to keep in touch and then rowed back to our boat.

Next day, as we hauled in the anchor, Jerry and his son JJ rowed over and presented us with a fine bottle of champagne. This wonderful send-off gesture was followed by the realisation of how fortunate we had been to have met up initially with Jerry, Randi and their family back in San Francisco. We had experienced many enjoyable times with them that had very much enhanced our travels.

CHAPTER 10

North Pacific to Glacier Bay

"There was a mass of people on the deck filming and waving. It was quite an emotional moment for us"

We departed Hanalei Bay on June 27, 1978. We did not fight the northeast winds, anticipating that in a couple of days we would be picking up westerlies. Heather's dad, when marking our start on his chart, was wondering if we were headed for Japan!

Pilot Books are always interesting reading, but frequently dire warnings might induce one to stay home and not attempt to sail in this area. The North Pacific Pilot was no exception and was good if not scary bedtime reading:

> "The Aleutian Low looms over the North Pacific as a climatic warning to mariners navigating the Alaskan waters. This semi-permanent feature is made up of the day-to-day storms that traverse these seas in a seemingly endless procession. And with these storms come the rain, sleet, snow, the howling winds, and the mountainous seas that make the northern Gulf of Alaska and the southern Bering Strait among the most treacherous winter waters in the Northern Hemisphere . . . sustained winds may reach 60-70 knots . . . extreme wave heights of 60 to 75 feet."

The description was enough to make one throw up and forget the whole trip. Fortunately all the above occurs in the winter months. Our summer trip was to be one of near perfect sailing in more or less calm seas. This was in line with the experiences we had gleaned from two or three of the few people we'd met who had sailed this route.

We still made our ham radio calls to Vancouver and to other yachties. Our celestial navigation continued and Jeremy was now taking sun shots and computing the daily position. This invaluable experience allowed him to navigate a race boat in the Victoria-Maui race and later skipper the boat back for the owner. School work was moving along at a good pace and Jeremy and Erica were on schedule with their assignments; this was a fine achievement because on return they could join classes where their friends had advanced to.

It soon got colder and the log said on July 2, *"Long underwear on!!"* I wondered where they had been stored throughout our trip. Erica remarked that some dolphins that visited us did not have dorsal fins. The transmission fluid was down a quart. Some ships were seen. Jeremy was doing an assignment on Bertrand Russell. We experienced foggy conditions and then beautiful sunny conditions. Heard Borg won Wimbledon in three sets. Jeremy suddenly emerged from the aft cabin hatch with a great shout dressed in long johns, braces and a sun hat scaring us and raced all over the boat ending up in the rigging. Oh! The madness of being at sea!

From my daily notes:

> "July 5 – Some excitement tonight while on the ham net as we saw a 4' by 4'sun fish floating on the port side with its large vertical fin flapping to and fro in the air. Does this thing really propel itself? A small shark clearly seen below us in the water was sniffing around this odd creature. Twice now on my watch I have heard dolphins and seen their whitish phosphorescence trails zig-zagging in the water. Erica and Heather have made out that I was going loco since I seemed to be the only one to see them. So I woke them up one night to come on deck to see for themselves and prove I'm not losing it.
>
> A spectacular event today! We had a wonderful and incredible display for about four hours from hundreds of dolphins surrounding us and stretching away into the distance in all directions as far as we could see. They must be a pretty formidable and fearsome sight to any smaller fish as they would make for very efficient fishing, hunting in packs like that.
>
> The waters are so clear that lying in the bow while we are motoring we can see them at least 20' down racing along with us with occasionally one shooting past in the opposite direction. We've noticed that they always surface on the front side of any wave presumably to get the benefit of its forward motion. Sometimes it seemed half the pack would take off in another direction, presumably for fish, and then all come racing back plunging and leaping often 5' or more out of the water until they caught up with us again. It was an incredible sight to which I devoted a whole roll of film.

We have also had a booby bird following us lately with its large hanging wings gracefully speeding in and out of the wave tops with its wing tips just brushing the water; a masterpiece of flying efficiency as they rarely flap their wings seemingly getting all their lift from the motion of the waves. This bird had been landing in front of the boat and as it spread its wings up for breaking, its webbed feet stretched out behind, perhaps for balance, and at the moment of landing it whipped them forward, skidded along the water prior to setting down. During take-off it would start a running flapping motion for about 20' and go straight into its gliding motion again. We never saw one take anything from the water.

Today Erica has been writing to her friend, Karen, who is going to join us on Vancouver Island. I noticed she had a long list under "What to Bring" so suggested she limit it to five suitcases.

At latitude 40 degrees it was starting to get chilly and now at 43 degrees it is quite cold. My hands are cold writing this on watch. Our protective waterproof sheet, rigged up over half of the port side of the cockpit to keep the spray off and cut down the wind chill, is doing its job well; we do not have an enclosed feeling and have clear views of the ocean expanse. Porridge has been a must at breakfast for a good start to the day and we are eating more noodles, potatoes, cakes, etc. in the colder climes. At night we have multiple layers of clothes on, plus wet gear because in the fogs, that exist daily now, surfaces get very wet. Jeremy noticed his breath for the first time and Heather made up a hot water bottle after her watch. She has also noticed more comments in the logs as "Freezing", "Think it will snow tomorrow", "Steam from breath" . . . Hey! Where did those lovely warm soft breezes go?

We are moving along at some 5-6 knots with the big genny poled out and the 110% lapper working as an inner foresail. With only six days to go I feel I am straining to get there to the shelter and serenity of the mainland islands before the current calm seas and weather deteriorate. We are headed for the Canadian weather ship located at "Station Papa", 145 W/47 N, 862 miles from Victoria, which we should reach in a couple of days. I wonder if the ship maintains its position by going around in circles?

Maintenance never stops on a boat. This time the starter only just got the engine going although the batteries were showing to be reasonably charged. I manage to take the end off the starter without moving the whole thing. Notice one brush worn right down and sticking. Fix that and thoroughly clean armature. Starting seems OK again at least for a while. I must radio Klassen Diesel to have a couple of brushes flown up to Pelican. I cannot understand why there are so many problems with a marine starter at 800 hours use when our old Rambler station wagon had done 99,000 miles – and sold for $500!

Our frequent bilge checks show a lot of water and oil. The water is coming in through the 25 Whale Gusher pump when the backflow rubber flap gets a build-up of dirt on it. I cleaned it for the second time on the trip. There is a natural siphon since the bilge is 3' below the water level outside. I should get a siphon-break installed. The oil is another problem and I find it coming from the output end of the transmission box and spraying the engine room; room is a generous term as there is barely room to squat. Heck, I just paid to have the transmission rebuilt! I think consumers need a lot more recourse to corrective action by suppliers but Hilo is a long way back now! The refill tells me we have lost a pint in 50 hours running. Fortunately I had spare transmission fluid and engine oil.

At these latitudes the nights are short and at 3:30am I can almost read this. I could if there was no fog. Jeepers! It is getting cold and my nose is running. I carry a handkerchief now for the first time since we left Mexico! With this calm weather Jeremy and Erica are really getting down to their school courses. Erica has not enjoyed today but has done two math papers. Jeremy has been grinding along with "A Separate Peace" and tomorrow we will be involved in Richard III. Heather is becoming an expert on Shakespeare and I get questioned on math or where help is needed.

We have motored 50 hours on this leg so far and this is a chore to stand and helm as our Tiller Master has packed up. After three or four days of flying the chute I packed that in too because it was just too much strain standing and steering for 8 hours a day. So I reckon we'll lose about 10 miles a day. I can stand that loss for the freedom our reliable windvane, Goldfinger, provides. Jeremy is disappointed with me and believes I am losing my racing instincts. I told him it's that or falling arches. Downwind and reaching, Goldfinger is working perfectly ranging +/- 5 degrees off our course but does not work too well with the chute up and moving around. Heather thinks it is because he is just jealous of the red and yellow chute and goes on strike. Writing a log is a good way to pass a watch and I wish I had started a lot earlier but generally there has been too much motion. At the moment we are just gliding over calm seas and it is difficult to believe that in the winter this can be a cruel place.

Tuesday July 11th – Spoke to Restless Wind on the 20 metre band finding they had motored all night on their way to Seattle and were just getting into some wind. I compose a poem for Heather for our 21st anniversary and consult with Erica who is "secretly" baking a cake. Watch the sea build steadily from the south west. In the afternoon boat is slewing too much for Goldfinger so put a third reef in the main. As night falls I drop the 135% genny and pole out the 110% lapper. Still doing 6 knots. Hear through the DDD net from John Savage on Encourager, that the weather for us is 14' seas and 25 knots wind south west at about 50 degrees N and west of 140 degrees west. Thanks! We are at 49-10'N and 146'W! In anticipation of rougher weather, I hank on the storm jib and sheets and get all objects such as books and sewing machine tied down. We are 80–90 miles south west of the weather ship which we hope to pass in daylight in the morning. Will start calling them soon and see if they can spot us on their radar. 650 miles to go to Pelican.

Wednesday July 12th – Great excitement today as we look out for the weather ship. Our course should take us about 5–8 miles to the east of their position as we do not wish to get too close at night. Tried VHF at watch change and at 0900 hrs had a reply. Explained who we were and asked about the weather and whether they could spot us on their radar to which they replied no. Their replies seemed pretty short with no questions asked of us and we felt a bit as if we were the fourth #29 bus to pass that morning! So we just said that we would be on standby and would call in a bit later for a final weather check. The crew looked a bit disappointed. About an hour later they called and the captain came on and said they had spotted us on their radar. He gave us our position as 49-52'N and 146-50.5'W and advised it was accurate to within 400–500 feet. I like that as it put us two miles off our running fix made earlier."

This time we had a fine chat with Capt. Randy Dykes on

the 400' weather vessel *Quadra* from Esquimalt with a complement of 73 men and 7 women aboard. I asked him if he had seen a Swiss boat that had left two weeks earlier than us (for Alaska) and he said no, we were the first sailboat they had seen in three years! We wondered about the somewhat unresponsive reception we had had an hour ago. The boat they had seen was John Guzwell on his boat *Trekka*. John was one of several people who had recommended we take this route home. We had met him and his wife Maureen at the home of Charlie and Liv Kennedy when starting to build our boat.

We told Capt. Dykes about the DDD net and went on to discuss a number of subjects including the original version of Beaufort wind scale. This scale compared wind speeds to their effect on a sailing man-of-war. He said they would come down off their station and visit us. This generated all kinds of activity aboard *Sky One Hundred* including running up the Canadian flag, taking down the blue cockpit cover and Erica combing her hair for the first time in many days. Actually, to her horror, knotted pieces had to be cut out with scissors.

As we scanned the far distance, a white golf ball slowly edged its way over the horizon followed by a brilliant red hull. Quite suddenly it seemed the *Quadra* was roaring alongside with its close massiveness so scaring me that I altered course. We seemed about 150' apart! There was a mass of people on the deck filming and waving. It was quite an emotional moment for us, out in the middle of nowhere, with this huge boat alongside.

We were invited to come aboard and use their showers and have breakfast but I declined when I thought of the hassle of one of their boats coming alongside and perhaps damaging our boat. We went on to talk about the number of salmon they caught (some 500–600 per trip); the worst storm they had had (60' waves apparently once in a hurricane); and, the original wording of the *Beaufort* wind scale again. This scale, created in 1806, assessed the wind force in terms of the amount of sail a man-o-war could comfortably carry without damage. At Force 12, the strongest force, the description was, *"Hurricane, or that which no canvas can withstand"*. I came across this wind scale when working on a project in India.

Capt. Dykes was retiring after four more trips (six weeks on and four off) and was a very pleasant person. Then he said there was someone aboard who knew us and wanted to talk. On came Bev Oakley from our tennis club. She had been at Jeremy's school and was aboard on a training course. Oh! receding world! She and her father had beaten Heather and me in a doubles competition. I kidded her that they had won by some devious play.

Capt. Dykes advised that they were leaving us and wished us safe sailing. I thanked him for the special thrill of his visit and for the offer to send pictures of us under sail. The *Quadra* then turned up wind, released one of its weather balloons and, as we watched, she slowly slid back over the horizon leaving an empty flat rolling ocean. The world is round . . . .

> *"Thursday July 13th – Came on watch at 0200 and find we are moving along at 5-6 knots in calm seas. Today is our wedding anniversary and Heather and I have a special hug – in fact there are hugs all round. Jeremy produces a poem for us and Erica brings out an iced cake with "21" on it for us. It has a swimming pool look as it is deep at one end and shallow at the other due to the heel of the boat and because we do not have a gimballed stove. It is gobbled up in a remarkably short instance! Discuss some opening scenes of Henry I with Jeremy and think I could get back into Shakespeare now. Call up Helen and Chris that night on DDD and Jerry Anscombe finally locates them at the Sidneysmiths' house. Have a great chat and invite them to come up for a week and visit the glaciers and Skagway.*

They will consider. Heather and I have an extra stiff drink that night while ignoring my watch.

- *Friday July 14th – We find that the current is carrying us north of the circle route so head down 5 degrees to counter. Talk to Restless Wind and Encourager and find they are only doing 60-80 miles a day and reckon 10 days more to Seattle. Also spoke to Tyee in Seattle who wondered if we knew where Free Spirit was as their ETA was today. They had left Kauai two days before us. Reckon we have about 300 miles to go and should reach Pelican, population 133, on Sunday. Hope there is no charge by Customs. At this latitude of 54 degrees I note that 1 degree of longitude is about 34 miles.*

Heather saw two ships in the night so believe that we have passed through the Yokohama to Vancouver shipping lane. She says she came and called me but that I fell back to sleep. Oh! These late nights! Normally, if the watch sees a ship at night, they were to get a second crew member to come and confirm which way a ship is travelling as sometimes one can get confused on exactly the ship's direction. Lost the topping lift up the mast this morning. OK, as we still can use the spinnaker halyard. In spite of Capt. Dykes' advice on fishing lures, we still do not catch a salmon – must buy some decent lures. Champing at the bit today as our speed is slowing and ETA at coast is sliding back to Sunday evening which is not good as Pelican is 3 hours further on. Will throw up the chute early tomorrow. I think we are all getting, not bored, but ready for some continuous sleep in quiet anchorages as we seem to be slowing down a bit ourselves.

North Pacific to Glacier Bay

"The whole vista was magnificent with partly snow covered mountains marching away down to the south"

Hear on DDD net that Whitecaps soccer match had 3,000 at their game in Vancouver and that the temperature there is 78 degrees. See some kelp for the first time – must be getting close to land.

- Saturday July 15th – Find mistake in DR or is it the current effect or me being sloppy? Heather and Jeremy getting tripenditus. Still very calm with wind less than 8 knots from WNE. Use spinnaker with motor ticking over to roll along at 5 knots. Calculate ETA Monday at 1500! Erica is sick from a bad egg. I sleep enormously. Heather bakes and we have afternoon music and drink. Can receive Sitka radio now.

- Sunday July 16th – After wind and rain all night both stop at 0500 so we motor all day finally introducing the Tiller Master which I've repaired and now works OK except that it corrects boat movement too quickly and does not allow boat time to roll back and forth in its natural movement. I'm a bit edgy today as the ETA is sliding back and last miles are dragging. ETA now Tuesday at 0600.

Sight land at last – Mt. Edgecombe, a 3,500' volcano, 60 miles off and on the correct bearing. This leaves us about 55 miles to go. I'm sure I can smell fir trees. A beautiful clear day is developing along with a flat calm and a slight swell. Lots of jelly fish are appearing. They have a filigree web with a red dot in centre. Some have four webs together. Last night there was a big bang and our fish hook was gone. Notice little vees in the water made by something small that moves when our bow wave moves out. Cannot get the DDD net as too much static and QRM (interference).

Perhaps no one has their beam up here. Will ask John on Encourager to relay message tomorrow.

There is a fantastic lilac-coloured sunset. In fact over the last few evenings there have been some superb colours in the sky, variations of mauves to reds which, reflected in the calm seas, have created unusual and fascinating vistas. There will be a moon tonight and I will shoot Polaris. Cannot see any lights ashore yet. Transmission seems OK but exhaust line is leaking somewhere and sending out rotten fumes. Toilet is definitely worse and needs a thorough fresh water cleanout.

Monday, July 17th – Today is Jeremy's 17th birthday and our 20th day at sea. He is suitably congratulated and another cake has been prepared. I write a poem for Jeremy. He is such a super and competent fellow. If only he would not take so long to come on watch! Ha! Ha!

We have been out one year now and Heather bakes a steam pudding, my favourite! I am such a lucky guy. I can definitely smell trees now.

At 0245 hrs Hawaiian time (local 0545) the sun just tips over the mountains and streams across the rippling mirror of a sea. A multitude of snow-covered mountains rise above the lower coastal mountains. Mt. Edgecombe with its symmetrical cone shape looks impressive to the south east with streams of snow radiating away from the crater.

Passing by Hill Island,(was this named for us?) we find picking up marker buoy for Lisianski Strait difficult to see in spite of deliberately aiming for the south side so that we would definitely know which way to search along the coast for the entrance. The problem is that as one nears the coast the prominent landmarks merge into the general terrain while the continuous expanse of trees makes it difficult to distinguish a narrow entrance. We make a general request on Channel 16 and a fish boat replies saying they are just about to emerge from the entrance. We see them coming from behind a small headland. We also see two whales leaping and cavorting about half a mile off – are they welcoming us?

It is interesting to note that in 1741 two Russian vessels, the St. Paul and St. Peter, from Kamchatka, under the command of Vitus Bering, were sent to find and explore America. While starting together they became separated but reached this coast separately. The St. Paul, when off the Lisianski Inlet, sent a boat ashore with ten armed men. After six days it did not return. A second boat was sent ashore but it did not return. Natives were seen but the Captain suspected the worst and returned to Kamchatka. Meanwhile Commander Bering on St. Peter, dropped anchor off Kayak Island (we were to anchor there in 1984 when sailing from Prince William Sound to Glacier Bay) and allowed his scientist, Steller, four hours' exploration ashore before returning to Kamchatka. I ask myself why such a short stay after such a huge effort to make the trip. Perhaps winter was coming on. On the return the boat was wrecked on an island where Bering and many of his crew died. The remaining crew built a small boat from the remains of St. Peter and managed to return to their home port. It is said that Steller then returned by horseback to Moscow from whence he had come from for the expedition in the first place! They must have been exciting and potentially dangerous trips in themselves!

It was 37 years later before the next white person was to arrive on this coast. Capt. Cook arrived in Prince William Sound in 1778.

> *"There was no sound, no movement, no surf.
> What utter bliss!"*

We are getting very excited now. The whole vista is magnificent with partly snow-covered mountains marching away down to the south to Mt. Edgecombe and the horizon. There are brilliantly snow-white covered mountains looming in the far distance hinting of some massive ranges we may yet encounter.

We pass over an 8 fathom line and then back into a 100 fathoms plus as we head up the Lisianski Strait. Perhaps the change in depth resulted from a tidewater glacier ten thousand years ago being static on the entrance and dropping its moraine. We chat on VHF with Lady Grey, the fish boat."

WE ARE BACK!!

We did skins all round. A tremendous sense of relief swept over me. I felt a lightness and an exuberance as if I had just passed a major examination. I now knew that unless we did something rather stupid we were off the high seas and safe from any major storms. This worry had probably been buried inherently and unrecognized within my system and waiting for some release. I rushed onto the foredeck whooping loudly and hung off the bow over the water while holding onto the forestay and waving an arm and a leg over the side. We all shouted at passing rock faces and received distorted echoes back. The challenge of building *Sky One Hundred*, planning for this trip and getting to this point in a safe manner was over.

I went below for a catnap. I was woken and told oil pressure had dropped. I checked oil level and found it too low. Now what? We only had one quart left and 15 miles to go.

We threw up spinnaker and sailed for an hour. Wind dropped so we motored on again and powered for Pelican.

We arrived at 1600 hours and checked in with Customs located in a dilapidated shed – no one there! Checked in at Post Office and learnt the lady customs officer was in Las Vegas. We suggested she was living it up but no, her mother was seriously ill. We asked for mail for Hill and were advised there was none. Suggested that the parcel (our new starter brushes) on shelf behind counter which clearly had our name on it was ours and it was handed over. Shades of the Marquesas!

Pelican was an isolated settlement. The village clung on the mountain side above the sea and beach. The main thoroughfare was a 15' wide wood boardwalk on trestles about 20' above the sea. The tidal range was about 28' which could make for some big currents. All the buildings were timber and there were wooden steps leading off to terraced houses behind. It was a very quaint and tranquil place. As we shopped and did our laundry the locals seemed young. At the well-stocked food and marine stores there was just $11 dollars left from our $100 bill. Phew! Bread here was $1.15 and in Seattle $0.59. We never enquired about steam baths.

We bought oil for the engine and motored on up to Mike Cove. We called up the net to talk to our Vancouver neighbour, Joan Lovegrove, to ask about Heather's parents' departure. Transmission was not good and Jerry who was running the DDD net today advised Helen and Chris had left for Alaska. The transmission faded with static and QRM and I heard Harold Schnetzler in Hawaii trying to relay for us.

Mike Cove was beautiful and calm and I was ecstatic to be there and in the northwest Pacific again. We were amazed at the extraordinary lush green-ness of stunted growth above the tree line, like English greens. Higher still there was raw rock leading to patches of snow. At anchor it was so quiet and still as we lay abed. There was no sound, no movement, no surf. What utter bliss!

Next day we rose late, up-anchored and breakfasted on the way to Bartlett Cove at Glacier Bay to meet Helen and Chris for the first time in a year. We were all very excited. We cleaned up the boat and ourselves while Heather baked a special cake. Jeremy started a score card for eagles seen, already 21. The mountains west of Glacier Bay were the Fairweather Range and were clear and spectacular. Mt. Crillon, 12,727', together with Mt. Berthon, 10,182', each with a cap of snow and ice, stood majestically white and stark above the lower lying darker hills and the grand sweep of Brady Glacier which flowed down to the sea. The whole scene provided a chilling reminder of man's fragility.

We passed through South Indian Passage and were in sight of Bartlett Lodge so called on the VHF. The ranger station answered and we asked them to pass a message on to Helen and Chris at the lodge that we will be there between 1500 and 1600 hours. I then called John on *Encourager* on our ham radio who was en route for Neah Bay from Hawaii. He answered immediately (bloody amazing!) and I told him this is a fantastic place and wished that they and *Restless Wind* were with us. He would relay a message to them and the DDD net with our good wishes. This instantaneous communication system floored me and I went on deck and shouted with glee. We took photos over the bow of the family's reflection in the smooth mirror-like water.

> *"The desolation and rawness had a very powerful effect on me"*

With less than an hour to go, more mountains and glaciers appeared as we headed into Icy Strait. A pod of whales started blowing and leaping from the water. Dolphins slid by and a ketch was coming up behind us. The lodge dock was now in sight and we could see Helen and Chris there complete with sleeping bags and cases and waving wildly. We docked alongside the tour boat. There were a lot of emotional embraces and tears with everyone talking at once. Finally we settled down in the cockpit and started catching up on our twelve-month absence. Later Howard and Norka Elting, who'd met Helen and Chris at the lodge, came over and joined us for drinks. In no time Howard and Chris were quoting Shakespeare to each other.

We had supper at the lodge where the glasses were filthy but the food was good. Howard and Norka sent over a congratulatory bottle of Pinot Noir wine, a nice gesture. Later we moved our boat and anchored out. We were up at 0530 hrs woken by the frenzied whine of huge mosquitoes. I checked the engine and transmission oil levels and we departed by 0600 hrs. It was rather overcast but no rain and unfortunately no wind.

As we moved into Glacier Bay we noticed a sleek white cruise ship coming up astern and to our immense surprise recognized it as the *Sun Princess* last seen and talked to off Cabo San Lucas in Mexico. We called up David Lumb and said we never expected that they would personally come and escort us into Glacier Bay. He wanted to know where we had been. Later we talked again to him and the captain and they wanted us to come aboard for lunch at Margarie Glacier. We regretfully suggested we will be travelling too slowly for this but hoped to meet up with them on their next trip at Skagway and if not then, certainly back in Vancouver. They speeded off into the distance as we seemed to be bucking a strong current.

As we watched some more whales blowing, leaping and cavorting a half mile off we noticed another cruise liner coming up. We realized it was the *Veracruz*. We hoped friends from our Capilano Tennis Club might be aboard but we could not make contact on the VHF.

At Willoughby Is. we started seeing the first ice bergy bits in the sea, until at 43 miles up Muir Inlet from Bartlett Cove and 2 miles from the face of Muir Glacier, we could go no further; it was impossible to work our way through the spread of all sizes of bergy bits across the sea. It could also be dangerous if the wind started to blow them in our direction.

The whole scene was wild and primeval with rock faces scraped clean of any soil and vegetation. Great U-shaped valleys ceased abruptly and hung into space. Lateral moraine banks covered with new deciduous growth lined the channel we were traversing, broken through here and there by the passage of past secondary streams of ice. Behind, were the dark green of fir trees and above that, the lush tropical green of moss and bushes frequently right to the top of a mountain. In areas where the ice had receded from (less than 200 years ago) the rock was bare with striation lines sometimes clearly seen.

The excellent Glacier Park information pamphlet ("The Land and the Silence") explained that when Capt. Cook came here the ice was down past Bartlett Cove and into Icy Strait. Since then it had retreated some 40-50 miles. The most rapid rate of retreat has been 5 miles in 7 years and it is still receding. What will be left for our children's children to see? While there is a current concern about global warming it surely has to be realised that there was a massive ice sheet over Canada only some 10,000 years ago, so warming problems surely cannot be laid solely at mankind's feet.

At either side of the inlet we saw fresh tongues of ice, including the spectacular sweep of Casement Glacier with its three lines of moraine dirt disappearing right into the clouds. We wondered if each line represented a glacier joining higher up. The lower levels of the glaciers looked rather forlorn with ice receded and shrunken from the rocks and covered in dirt. This park is known for its surging glaciers. One was said to move 1,100' in 24 hours – we could not conceive the noise, vibrations or wave that would have occurred. Normally movement was much more stately but there were variations of the same glacier as, at one visit there may be a sea of bergy bits, and at another time one could get right to the glacier face.

We noticed where there were bergy bits at glacier faces there were many seals resting on them. It was said to be a pupping area. Perhaps the area was attractive to fish. Apparently Indians used to camp near the face of glaciers as the seals would be a convenient food source.

As the cruise ship *Veracruz* was close by, we made contact asking if our friends were on board. They were and on came Ray and we had a wild chat on the VHF radio for a while ending with good wishes for a safe and sober trip! We grabbed our tennis rackets, raced onto the foredeck and started waving wildly and shouting as we could see them waving as they went past. It seemed the size of the world was much dependent on the number of people one knew – or more technically, was inversely proportional to the number of people one knew!

Another centre cockpit boat *Sun Piper* from Seattle caught up with us and had turned. We took mutual photos and exchanged addresses. The ice was quite thick again and I winched Jeremy to the top of the mast to take movies down on the boat showing the ice chunks (3–20 feet across) sliding past our hull. We passed *Sun Princess* on the way down the inlet and David Lumb advised we could get much closer to the face of Margarie Glacier so we decided to go there the following day. We anchored close to Lamplugh Glacier alongside *Sun Piper*.

It was perfectly calm and peaceful except for the chirping

and squawking of birds on a small island which hung suspended in equal greys of low fog and water; there was no horizon as they merged. Faint whites of ice floes drifted by. It was difficult to imagine this place may, in the winter, have winds over 50 knots, 10' seas and freezing spray.

We moved off at 0600 hrs and headed up Tarr Inlet sticking close to land. The engine was overheating so I added fresh water and two quarts of oil. Later the mist lifted and we had a superb day arriving at Margarie and Grand Pacific Glaciers in the afternoon. Chilly drafts swept down from the glaciers. We were alone in this totally desolate spot. There were far fewer floes here and we could get right to the face of the glaciers but remained 100 yards off. We sat with the motor off and listened to the continuous rumble like heavy gunfire and the crash of ice. There was some calving of ice from Margarie Glacier and at each crash we tried to see where to get some photos. We basked in the sun as the blue of the sky increased. We were very careful motoring through any fields of bergy bits as a small 3' cube could make a big bang on the hull or worse, damage our propeller. We lassoed a fair sized ice floe and towed it for a while; we were perhaps one of the few boats which had done this in spite of some scientists' wild plans to tow huge bergs to the Middle East to provide countries there with a new water source.

While we were in American waters we noted that the Grand Pacific glacier face had receded north close to the border and one might soon be able to sail into Canadian waters. This whole wonderful park was a fascinating area encompassing massive tidewater glaciers, perhaps the last of some of nature's giants. The desolation and rawness had a very powerful effect on me. I know I will want to come here again. It is said that one should not visit Alaska until you are old because other vistas will pale in comparison; I can believe that.

The park pamphlet indicated that Lituya Bay, just a few

miles outside of Icy Straits, had in 1958, a massive rock-fall caused by an earthquake of magnitude 8.3. The bay is T-shaped and about 8 kms in to the top of the 'T' where the Fairweather fault line exists. Two glaciers exist on either side of the 'T'. The effect of the rock-fall was to drive a wave of water 1,720' up the mountain opposite, nearly half the height of the North Vancouver mountains! The British Columbia Centennial climbing team had just made a successful ascent of the 15,300' Mt. Fairweather and were celebrating on the beach when their pilot advised he was coming to pick them up a day early as the weather was about to deteriorate. They had just been flown out of the bay when several hours later the earthquake occurred and the wave swept across the bay! It caused massive damage sweeping three fish boats over the spit and into the sea where one was lost. The sea was covered with trees for miles around. Our friend, Kelly Duncan, was the cameraman for the expedition which included Paddy Sherman. When we sailed into Lituya Bay in 1984 we could see the new growth of trees up to the 1,700' level.

It has been estimated that massive landslides into the sea can create *super* tsunamis which can cause terrible damage when they hit the shore. Such a wave might occur if a certain volcano in the Canary Islands erupted and caused a major rock slide into the sea. The *super* tsunami would cross the Atlantic and badly damage the east coast of the United States.

Close to Lituya Bay it was interesting to note the huge La Perouse Glacier was most likely the last glacier still actually touching the Pacific Ocean and directly facing Hawaii; the rest having all receded back up channels.

We managed to pinpoint Mt. Fairweather by checking the height with our sextant at 15,200' which was 100' off its recorded height. An adjacent mountain was 2,000' lower. After two hours we motored back to John Hopkins Inlet. It was a beautiful evening with the sun setting behind the mountains so I got into our dinghy with the movie camera while the family took *Sky One Hundred* back a ½ mile, hoisted the spinnaker and then roared past me with a glacier in the background. Fortunately they remembered to pick me up!

We dropped anchor in 40' on the side of Tarr Inlet. There being no other sheltered anchorage for 20 miles, another boat anchored with us. They wondered if it was safe and we suggested it was. While exposed to big fetches and with occasional bergy bits flowing by we found it, after our offshore anchorages, to be safe enough. If the wind blew up we would just up-anchor and start sailing.

It was quite calm and we dinghied ashore and climbed up to the Lamplugh Glacier. We collected in our net probably what was 10,000 year old ice for our GATS. That number of years ago Vancouver was covered in 5,000' of ice. Its quality was measured by the clarity with which we could read our winch name through the ice. Next day we motored to the glacier's green-white face and took more photos and went ashore. We noticed how slippery and fine the mud silt was at the face, presumably a result of the centuries of grinding of the moraine by the ice. On a recent visit (2013) to this glacier the face appeared to be in the same location after 35 years.

On returning to our boat it was up with the spinnaker for a fast run down to Bartlett Cove trying to arrive before the fog socked in completely. We made it before dark and were very relieved and fortunate to have 'done' Glacier Bay in good weather.

Next day we fuelled up and left in rain and mist; a fish boat 50 yards away could barely be seen. After a fight with the engine going full blast and the sails up we just, and only just, managed to beat the current and round Pt. Gustavus. Three hours and 10 miles later we were still bucking the current on the south side of Pleasant Is. Unfortunately it was too rolly for Helen if we anchored so we motor-sailed across Icy

Strait and into Flynn Cove on Chichagof Island by 2300 hrs and anchored. The mist was still on and off.

We were up early and motored up Chatham Strait for three hours and as the mist lifted we could see clouds racing up Lynn Canal leading to Skagway, our destination. Great news and we raised the spinnaker, put Helen and Chris into the forward bunk for a 'kip' and roared 70 miles up to Haines in about 10 hours.

The mountains on either side were spectacular, the Davidson Glacier very impressive with the evening sun back-lighting it against the dark of adjacent slopes. The mountains behind Haines were superb but the customs man took an hour or two to fill out the forms. We were very tired as he went through the process. We were only the second boat to come from offshore in three years. As he painfully filled up the form (who on earth cared what was in them?) he volunteered that 2,000-3,000 eagles could be seen feeding on the salmon in the spring; that the Chilkoot Valley supported some 1,500 black bears and 200 grizzlies; that five years ago 73 cruise ships came there and now there were none. What with that, a poor fishing season and too high a cost for timber, Haines was dying with people leaving daily. As he completed the last question in the form we collectively sighed with relief only to realize in utter disbelief that he had not put in his carbon sheet and would have to copy it all again.

From Haines it was a short step to Skagway. After a bit of dredging we entered a small marina and docked for the night. The next day the *Sun Princess* arrived and moored against a bleak bare dock which is backed by a high raw rock face covered with graffiti of different colors and signs. David Lumb invited us round for a CHAMPAGNE brunch.

When we reached the dock we found all the passengers were leaving to take the train up the Chilkoot Pass to Bennett Lake. At the top of the gangplank we were met and escorted to David's cabin, where there were introductions all round, and we sat down in real comfy seats and enjoyed the feeling of a big cabin that did not rock. Drinks were served with real ice, hey, this was the life! Actually drinks were served many times as we traded questions and experiences. The Captain and others dropped in from time to time for a look-see and a quick word.

After what seemed hours we arose, I use this word loosely with regard to the senior members of my crew, and wended our way through various passages to lunch. The very palatial dining area was empty except for a couple of tables with tour agents. We were shown to a huge round table set to magnificence, each place having layers of cutlery and different glasses. There seemed to be a waiter for each of us. I wondered, and no doubt Heather also, what the heck were we doing on a sailboat? Jeremy and Erica's eyes were popping at the sight of a massive smorgasbord spread which we leapt into and then the wine started flowing. Two hours later with waiters attending our every need, this style of life was definitely slowing us down so David wisely decided we needed a tour of the ship! Several group photos later, we said our farewells with promises to visit again in Vancouver and for him to visit our house for a change of scenery. Chris, by this time, was floating about a foot off the ground while I towed him back to our boat where we all crashed out. The *Sun Princess* left that night.

Skagway was a quaint old town and the most northerly part of our trip at N59.27.00. Lots of touristy wood-shack shops and old two-storey buildings were to be found around in scattered streets. It was possible to visualize the wild scene in the Klondike gold rush of 1898 in the Yukon as people flocked from the south to search for gold, to buy supplies and then get to Dawson City. Heather said she could visualize the activities of Soapy Smith, the gangster and con artist and his gang, among all the rookies heading up to find gold.

We took the White Pass and Yukon train to Bennett Lake. It was a pleasant interlude from sailing as the track rose 3,000' from Skagway, wending its way around the sides of mountains, through tunnels and over large, rather skinny looking, trestles with spectacular scenic views of mountains. We looked down on the rugged Dead Horse Gulch where 3,000 packhorses had died in the rush. Gold seekers had to climb the Chilkoot Pass many times as the Royal Canadian Mounted Police would not let them into Canada without one ton of supplies. The climbers were so close together that if one stepped out of line for a rest it might take an hour to get back into the line because there was little or no space between each climber. Actually gold seekers came via many different routes, including through Dyea in the next valley, and overland routes from Alberta and southern B.C. It was said that many rookies en route turned back realising the difficulty of achieving their dream; when they did so they could not convince those still pushing north they would not make it. Such was the desire to get rich!

At Bennett Lake we disgorged to view a museum, an old rail station, before being given a real *"miners' meal of moose stew and beans, apple pie and cawfee"*. On our 2½ hour trip back, after all the trials I had had with our boat toilet, I was pleased to view the ultimate toilet, one where everything just dropped straight onto the narrow gauge track – so simple, no valves, no pumps and no jamming.

Our trip down to Juneau was uneventful except when we suddenly realized a strong current was sweeping us into a vertical rock face and the Tiller Master could not correct our course quickly enough. One can never relax at sea so we were quite attentive passing the infamous Vanderbilt Reef. Here in 1918, the *Princess Sofia* ran aground in bad weather. The captain declined an offer to help take people off saying he would do it the next day. Unfortunately, at night she slipped off the reef in an increasing storm, sinking immediately and losing the 385 passengers and crew. Ugh!

We worked our way down to Juneau through a very narrow and shallow channel barely marked with sticks and into a marina. We spent a couple of days there sightseeing while visiting the 'famous' Red Dog sawdust floored saloon and the Mendenhall Glacier. After such a short visit we were all sorry to see Helen and Chris onto the plane back to Vancouver. Still it would be only a few weeks before we would meet up again. The visit had been particularly good for Jeremy and Erica who had been able to describe their experiences to a very interested audience. Card games were often played late into the 'long' northerly days until yawning seemed to overtake us all.

In spite of calls in Hawaii from my office as to when I would be returning to work we still had 5-6 weeks to work our way down the inland waterways.

CHAPTER 11

Homeward Bound

"We were able to luxuriate in our own private baths made of plywood"

After the pleasure of having Helen and Chris aboard, their departure was something of a let-down. The exciting offshore ventures and visits to exotic islands were over and now we had to settle into another rhythm – that of making our way south through a myriad of channels and islands back to Vancouver. We had the feeling we were on the downhill slope leading to work again. Many cruisers had said, *"Why, after all the preparation, are you just going for a year? We have been going for four years."* While I'm sure Heather would like to have continued I think a year out of the business world was enough for me – one has to earn a living. Despite this slight 'downer' I was looking forward to our exploration of this historic coast and experiences as we travelled home.

From Juneau there were various route options south. We chose to cruise down Chatham Strait, anchoring first at Funter Bay on the northwest tip of Admiralty Island. This site was a result of a cruel re-settlement decision by the American government. In World War II the American army, as part of a scorched earth policy in case of a Japanese invasion, burnt all the homes of the Aleut people who lived on the Aleutian Islands. The people were shipped to this bay where they lived in an old cannery in disgraceful conditions throughout the war.

As we started on this route, we were quietly sailing along when I put the motor on as the knots were dropping. Imagine our surprise when a submerged grey whale, 20-30 feet off our starboard bow, almost alongside, decided to dive and with a flash of its fin disappeared in a swirl of water – what a thrill but not time enough to take a photograph. Had our motor woken it from a dream? The weather was excellent, not a cloud in the sky with the wind from the north, just perfect for a pleasant downwind run.

There were so many wonderful Russian names saturating the islands of this coast: Yakobi, Baranhof, Chicagof, Kruzof, Petersburg and Kuprenhof. This would have been a wonderful area to explore but it was a long way to Vancouver. With more time we would have liked to have made the trip through Peril Strait, north of Baranhof Island and down to Sitka. This was the old Russian capital which was often fought over by the Sitka Tlingits.

However, we continued south to Pavlov Harbour, on Chicagof Island. Here we hiked up a trail to watch salmon coming up stream to lay their eggs. Two small float planes arrived for a few hours' fishing. We marvelled at the thousands of salmon fighting their way against strong stream currents and up small, but powerful waterfalls. Salmon were leaping out of the water everywhere. At one waterfall, salmon were waiting in packs right across the stream, keeping up with the current and trying to leap up the descending waters. Many times they would drop back and then have to try again. With our movie camera, bought in Hawaii, and on slow motion, the action of a salmon in and out of the water, trying to make headway, was fascinating. There was a powerful motivation at work here to return to where they had started. Above the falls in calmer waters there were lines of these anadromous fish laying their eggs in a sand bed. It was a very impressive display.

With a very strong consensus decision we stopped on a long dock at Baranhof to visit the hot springs. We were the only boat there. This was to be a real highlight, and, while we were still socially acceptable, it would be our first bath in over a year.

There were natural open springs about a quarter mile away but we couldn't wait as, at the top of the dock there was a bath house where we were able to luxuriate in our own private baths made of plywood. As I stepped into my bath and sank down, submerging myself, I felt my body was being immediately re-vitalised. A continuous flow of hot water poured in removing any residual stresses and plenty of grime. Was this some form of Heaven as I wallowed and gasped with pleasure in this liquid heat? Judging by the adjacent shouts the others felt the same way. While there was no time limit, my bodily systems were steadily sapped of energy and all too soon I had to get out while I still had the will and strength. Returning to the real world again we all felt quite light walking back to the boat. What a pleasure and a bargain, at a dollar each, to be followed, of course, by a couple of GATS.

After a short sail across Chatham Strait and around Pt. Gardner at the south end of Admiralty Island we dropped anchor by the derelict cannery at Tyee. I did a quick sketch of the cannery from the boat before night fell.

Humpback whales are enormous weighing up to 30 tons and reaching 50' in length. Back in 1907 one of the best whaling stations was built at Tyee by a Norwegian American from San Francisco. I am always impressed by these early entrepreneurs who had the will, drive and enterprise to inject themselves into the wilderness with projects such as this. They even built their own boat for whaling. It was initially a thriving success. They extracted blubber to be used for fine oils and soap, bone for corsets and made fertilizer with what remained. Within a few years the whales, perhaps smarter, moved further afield and the industry was replaced by a fish processing plant. When we were there, only derelict buildings, docks and wharves were reminders of a past industry.

I was reminded that my mother had possessed whale bone corset. It actually saved her from a serious injury. One day, when she was at the kitchen sink, one of my older brothers, Tony, put his pellet air gun into her back saying, *"Hands up, money or your life!"* She replied, *"Don't be silly"*, whereupon he fired the pellet leaving her with *only* a large bruise thanks to her corset. Suffice to say, it was considerably less than the sum of those she inflicted on him!

As we were unable to obtain a detailed chart for the passage through Keku Strait (this appeared to be an interesting but narrow route), we went north of Kupreanof Island and anchored inside West Point at the entrance to Portage Bay.

"The effect upon us was rather eerie as if the noise was a protest of our being on this land"

The following morning we sailed down to Petersburg, a small vibrant fishing town of 3,000 people located at the north end of Mitkof Island. Here we caught the tide for a pleasant trip through Wrangell Narrows and a peaceful night just northwest of Deception Point. We were still seeing whales, dolphins and salmon everywhere as we sailed between Mitkof and Zarembo Island. In Wrangell Harbour we stocked up on a few supplies.

In these calmer waters Heather was writing her 9th and final Newsletter. It would be posted in Prince Rupert to our friend, Doreen Watson, who had kindly volunteered to distribute copies to friends interested to hear the 'latest'. The Newsletters were perhaps also the catalyst for friends to send welcome home news back to us.

From Zimovia Strait we made a slight detour to anchor at Humpback Bay. Ashore we took a trail through the woods to where the U.S. Forest Service had an observation point for watching bears catch salmon during runs. It was an incredible sight to see again the water black with fish as they fought their way upstream. But we didn't see any bears until we were on our way back to the boat when one crossed our path carrying a fish in its mouth. It was concerning that although salmon have apparently been spawning in these waters for some two million years, current runs were very much depleted.

We cruised down Seward Passage between Deer Island and the Tongass National Forest on the mainland, Ernest Sound and around into Clarence Strait. A fairly strong southeasterly was blowing and we slid into Meyers Chuck, on the north side of Cleveland Peninsula, for the night.

After our ocean sailing the daily progress down the narrow channels set a somewhat different pattern of daily living; quietness with a good night's sleep for one thing, extra helming, continuous attention to the charts, deciding our route and when and where to stop, watching out for wildlife and viewing the continually changing vistas. It was also the last chance for Jeremy and Erica to complete their assignments. Our focus was also very much on when we would be home.

The following day we had a rough motor sail finally passing Caamano Point, at the south end of Cleveland Peninsula and pushed into Tongass Narrows and into Ward Cove at Wacker on Revillagigedo Island, tying up to an old barge for the night. Next day we took a conducted tour of the local pulp mill before leaving to ghost through some mid-morning fog into Ketchikan of some 7,000 people. The fog lifted to display this delightful town and its buildings on the hillside above the history-laden waterfront boardwalk. We watched with interest log booms being manoeuvred alongside a freighter. Workers on the boom organized logs to be individually lifted aboard for export.

It was 22 days since we had arrived in Alaska. We planned to be back in Vancouver at the end of the month. My office was wanting to know a specific date of return to work so that I could be scheduled into a project. A day later we sailed across the US-Canadian border into Canadian waters, and on August 10th tied up at the Prince Rupert Y.C. The Canadian Customs official was friendly and asked virtually no questions – there was no request for that little green card given to us when we left Canada and which listed all our equipment. I liked that!

The foreign segment of our voyage was over.

Loading up with 32 gallons of diesel fuel we headed off south down the narrow Granville Channel, passing the north end of Gil Island. Here in March 2006, a B.C. ferry, the *Queen of the North* carrying 101 passengers, ran aground at night and sank after failing to make a required course change.

The channel was quite narrow at times being less than 500 metres at its narrowest point. An approaching freighter seemed twice the normal size and we moved to the side when one passed us. The 70 km-long channel was steep sided with trees all the way down to the water and many waterfalls descended from the cloud shrouded mountain tops.

We were travelling with favourable currents whenever possible and were making good mileage as we traversed the

Princess Royal Channel, stopping at Butedale, Klemtu, and Bella Bella. Pleasure boats were rarely seen on our trip down through this unique cruising area and after a few days we looked out for boats to meet up with for a chat. We did finally meet up, surprisingly, with *Tai Shan* from our West Vancouver Y.C. and were invited aboard for a very enjoyable supper; I don't think anybody stopped talking.

Soon we were into familiar waters at the north end of Vancouver Island. At Port Hardy, Karen, Erica's friend, joined us and Jeremy left to get back more quickly to his girlfriend. When he decided to leave, Heather and I tried gently to suggest he had been away for a long time and that the strength of past relationships do tend to wane. We hoped he would not be disappointed.

Once when sailing along we heard the 'phish, phish, phish' of whales blowing and then could see them coming up behind us. With *Sky One Hundred* right into their path, we doused all sails and stopped. We were rewarded by having about eight big and small killer whales pass within 30-40 yards of our stern. It was a whole pod and perhaps one of the two pods that circulate behind Vancouver Island. While we had seen whales earlier this was a very thrilling and rare sight being so close to us.

In the Broughton Archipelago, a marvellous and fairly remote sailing area, we stopped off at the deserted Kalukwees Indian reservation on Turnour Island. The settlement, situated behind a rare white sand beach, presented a sad and dismal picture of dilapidated wood houses and a rotting wharf. This depressing but haunting scene was emphasized in the dying light by the black openings of the empty windows and doors. There were no totem poles standing. One could visualize an active self-supporting tribe living there perhaps not so long ago but where had they gone?

This was one of several deserted settlements we had seen

while sailing. Another deserted Indian settlement was at Mamaliliculla, on Village Island. Here we edged *Sky One Hundred* carefully through rocks to tie up at an old dock. As we walked ashore through an overgrown path we were accompanied by the raucous cawing of ravens or crows, we did not know which; it gave us a rather eerie feeling, a protest perhaps of our being there. Here again this strategic site held a few old derelict wood houses and a well preserved totem pole. The still standing main timbers of the tribal lodge were 3' in diameter and decoratively carved. Scattered around were a few more personal items leaving one to believe people thought they might return.

We cruised on another 10 kms to Minstrel Island. Back in the early 1900s this was a thriving community of some 3,000 loggers, fishermen, and miners with hotels, churches and brothels. Now there was virtually nothing left but a coffee shop, a few houses and a dock. However, close by at Lagoon Cove on Cracroft Island existed the only boat slipway for miles around. It was run by the Sedgleys, an Australian and his wife. Four years earlier on our first trip on *Sky One Hundred* we had a prop problem and had a need to use the slipway. We had called him up on the VHF and he replied *"It's available and I'll be waiting on the dock for you."* As we turned into the cove in a gusty 20-knot wind we could see the slipway coming out of the water and him on the dock but could see no structure to tie up to. Holding our boat up into the wind, we called out, *"Where do we tie up?"* and he referred us to a lonely 5x5-inch wood post sticking 6–7 feet out of the water. I yelled back *"You must be kidding, are you insured?"* On receiving two *"No's"* and there being no alternative we tied our boat to it with considerable trepidation but were safely hauled up and out of the water. We stayed overnight while I fixed the prop. The next day he showed us a 'bundle' of snakes that were entwined together at the back of his garden.

We stopped off at Port Neville but it was packed with fish boats so went onto Blenkinsop Bay. Liv Kennedy, author of *Coastal Villages* and a friend of ours, was born in Port Neville. Her book presents a superb picture-illustrated history of the early coastal life of the lower west coast.

We continued our way south through Desolation Sound. This was an area that we had explored so many times and spent so many wonderful sailing holidays. Perhaps this was the environment in which the urge for our trip started. The Sound, located halfway behind Vancouver Island, was especially attractive because of its isolation and the warmer waters resulting from minimal tidal movement.

Finally, as we came under Lions Gate Bridge we were met by *North Winds Five* again (it was them who had trumpeted us off with conch shells when we departed from Vancouver) who led us ceremoniously into a berth at the Vancouver Rowing Club. We had a wonderful welcome party of friends on board. Doug Cook, one of the original designers of the Fraser 42 was there with his wife, Elaine. As early members of the newly formed Blue Water Cruising Association we were the first boat to return from a major trip.

I think in the last days of this trip we were being assailed by two conflicting feelings. On one hand, a certain impatience had been growing to get home, meet family and friends and take up "the overheads of life" again and, in contrast, we were realizing our wonderful and much planned adventure was nearing its end. We consoled ourselves that we had not only reached home safely but we were healthy and enriched with new and rare experiences none of us would ever forget.

I reflected that my past dream for a further challenge and new experiences had resulted in a year of outstanding adventures and situations. Our family had joined a small group of people who had the good fortune, chance and perhaps the wherewithal to make similar trips. Making such a journey produces a profound effect on one, that of having a better perspective of people, a greater maturity and a further understanding of life and enjoyment. It also confirmed that dreams, aligned with motivation and desire, can achieve many things.

I was indeed a very lucky person to have enjoyed our voyage as much as I did. But especially, and most fortunately, I always had the complete and unstinting support of Heather, with whom to share this adventure along with her enthusiasm and positive outlook throughout. How sad it would have been to have just remained a dream. Being very close to Jeremy and Erica and seeing them blossom and grow was a rare pleasure during the intimate relationship with my family.

We had now become "Doners" and perhaps, as our experiences fade with time, we would again be "Dreamers" and "Doers".

Meantime Erica wondered how her Mum would like being on her own in the house on Monday, two days away, when it would be back to new schools for Jeremy and herself and back to work for me!

Afterthoughts of our trip to the South Pacific

From Erica:
At the time I don't think I really understood what we were embarking upon; later, in my twenties, aboard a Club Med "sunset cruise" near Puerto Vallarta and looking out over the Pacific Ocean, I realized what an incredible journey my parents had undertaken. Now, on reading this book, the magnitude of it really hits home again! It is amazing having so many memories brought back to life again . . .

- Finding out my parents' fears and worries but not having been aware of any of them.
- Visits to opulent yacht clubs, Disneyland and especially riding mules down the Grand Canyon.
- Realizing Mexico held, and still holds, my strongest memories and feelings.
- The "forever" empty white sand beaches and later taking the dinghy out through the surf.
- The novelty of spending Xmas Day in a swimsuit and turkey and ham coming out of tins.
- Sighting the Marquesas, very dark, foreboding and

quite ominous, and not as I had imagined it.
- The very friendly, warm and welcoming local people we met throughout the trip.
- Being in the cockpit at night, listening to the waves and hearing the squeaking of the dolphins.
- Receiving post being one of the biggest pleasures that could be imagined.
- The horror of seeing the wreckage of a yacht up on the reef with a big hole in the side, stripped bare and wondering how it must have felt at the time it succumbed to the sea and reef. It never occurred to me that it would happen to us.
- Travelling down the Inside Passage being much like our summer holidays spent going up the west coast but much wilder.
- Particularly enjoying times when there was isolation and independence from everyone, preferring the freedom of being in places where I could explore on my own.

At the time I didn't appreciate it but the trip undoubtedly gave me a sense of independence and confidence and almost a need to be alone.

From Jeremy:

I am impressed that we managed to remain compatible as a family for so long in the small confines of our boat. We never had any confrontations let alone any serious problems.

The extent of wildlife in the oceans and sheltered waters was always very impressive, be it massive cavorting whales or minute, coloured fish darting in and out of the coral and the nature around the Alaskan glaciers.

A big fear was to come back to Vancouver and not graduate with my friends and have enough credits for University. There were often times when I felt I wanted to drop the school course because there were so many interesting distractions but thankfully I did not. Fortunately I was a little ahead before I left at the end of Grade 10 and my correspondence courses had, on comparison, a much higher content than those in Vancouver and provided enough credits for university. Although I was away from my school, friends and my home life, I was to find, on my return, that nothing had really changed; particularly, I had kept up with the school friends and graduated with them.

My time on the boat had produced some moments of boredom but also moments of great excitement such as steering for the best speed down the waves off the Oregon coast and when we were laid over on our side at night just out of Oahu. Great fun was racing in the Coronado 27 Nationals off St. Francis Y.C. in San Francisco; racing in the Honolulu/Molokai in the Morgan 42 which almost sank, being over-powered and having too much water in it; and generally getting away on occasions with other guys.

As we leap-frogged with *Restless Wind* and *Active Light* it was fun to see them again, swap stories, views on life and go exploring with them. Meeting with other boats was always interesting.

All in all it was a wonderful trip and an introduction as to how other cultures lived, certainly widening my view on life, knowledge and my level of maturity. It has provided an excellent background of support to my work in the marine business; an invaluable time away from school.

I still think of the trip on a regular basis, am amazed that my parents were in their forties during the trip and we were able to enjoy so many experiences together.

From Heather:
I can remember contemplating our return to "normal" life while coming down the Inside Passage – just wish I'd written

those thoughts down. I do know that I wasn't particularly excited about returning.

Our trip had been such a wonderful experience – there was a sense of accomplishment in that we'd built our own boat, which had taken us safely on four long passages of the Pacific and all the adventures we had ran through my head; the relief of making our first passage from Vancouver to San Francisco, the excitement of meeting and making new friends and sharing exotic anchorages together, planning the next stop, etc. After a year of not knowing exactly what would happen next, the thought of settling down again didn't really excite me. I loved being on the boat – everything was so simple and close at hand and, once into tropical waters, the need for a wardrobe was really quite unnecessary.

I feel very happy that we were able to complete this trip as a family. There was a suggestion at one time that perhaps Erica would return to Vancouver from Hawaii, knowing how miserable the trip to San Francisco had been. When this was suggested she was adamantly opposed to the idea – "like copping out" and the trip to Alaska turned out to be one of our smoothest passages.

Now, in retrospect, it seems almost like a wonderful dream. Patrick was fortunate to be able to obtain a leave of absence from his company and I doubt whether an employee would be as fortunate today. Jeremy and Erica didn't suffer from lack of schooling on returning and it's fun when we are able to get together to share our memories. Being aboard a boat has a very big advantage when travelling in that you're able to see and visit places that the majority of people will never have the opportunity to enjoy. It's quite hard to believe how far we travelled in such a relatively short time and yet I never felt that we hurried.

My favourite memories are of being anchored inside a reef in calm water, within swimming distance of the beach, surrounded by magnificent scenery and blue, clear water – paradise!

It's some years now since we ventured offshore and I shall be forever grateful that I married a man with a sense of adventure; he's taken me to places I never dreamt of seeing.

APPENDIX

Equipment

ANCHORS

- 50 lb plough with 60 ft 5/16" chain and 300 ft 5/8" 3-ply nylon line (new). The 5/16" chain had a 1700 lb breaking strength
- 35 lb plough with 60 ft 5/16" chain and 300 ft 5/8" 3-ply nylon line
- 25 lb Danforth with 30 ft 5/16" chain and 150 ft 5/8" braided nylon line
- Mustang anchor winch manual with 700 lb pull (found not powerful enough)

SAILS

- 10.5 oz 90 sq ft storm jib (new)
- 6.5 oz lapper (jib) with reefing line (should have been 8 oz)
- 8 oz low cut 135% genoa: 35% of the sail overlaps the mast
- 5 oz high cut 150% genoa
- 1.5 oz spinnaker

- 8.5 oz main with an added third reefing point since we carried no trysail. We had added wear panels at the spreader contact areas.

DODGER

We added a low level hard windscreen of teak with unbreakable tinted glass with a soft fold-down dodger attached to the top which when lifted up could cover 1/3 of the cockpit. It was a godsend.

WINDVANE

Commercial windvanes for steering the boat were very expensive. I had also read of many failures of their sophisticated parts. So we made our own one based on an article by William Orgg in the Sail magazine May 1971. I managed to simplify the design and build it for a tenth of the cost of a bought one. We called it Goldfinger because the vane was covered by Heather with some gold-looking material from Gold's Fabric store in Vancouver.

NAVIGATION

We had two plastic sextants (new), charts to San Francisco, stop watch, Nautical Almanac for 1977/78, H.O. 249 reduction Tables and an explanation of the system. A metal sextant bought in San Diego.

ELECTRONICS

- Log and knot meter – Seafarer depth sounder
- Ray Jefferson radio and direction finder (new)
- Yaesu FT 301 transceiver (ham radio) with WWV time and weather (new) bought in Seattle
- Tape recorder 6V110V (new)

- Ray Jefferson 55 channel VHF radio (new)

SAFETY GEAR

- 6 man Avon liferaft (new)
- Avon Redcrest inflatable dinghy (new) brought back from UK on a business trip. These inflatables have a big strong rowlock which is capable of holding a large wood oar. None of these little plastic oars which if they break cannot be replaced off shore.
- 400,000 candlepower spotlight (new)
- Personal safety harnesses (new)

SPARES

All sorts of batteries, bulbs, electrical wire, voltmeter, rope, 60 ft of 1x19 ss cable in case a shroud broke, StaLok eye and fork fittings, rubber tire for slowing the boat down in a storm, engine spares such as filters, engine drive plate, water pump, zinc anodes, tools, hacksaw, electric and hand drills, spare strips of aluminum and spare sheets of plywood. Most importantly in San Diego we bought a Honda generator to produce 110V for electric tools and also to start the engine if the batteries were run down.

FOOD AND WATER

We had 125 gallons in two integral tanks. At 2 gals/day for the crew of 4 we reckoned that we could last 60 days. Food we were going to buy in the U.S. as it would be cheaper.

FIRST AID

Our doctor friend, Paul Watson, had fixed us up with a multitude of medicines along with an instant course on some do's and don'ts, such as "do not hit the sciatic nerve when giving a jab in the seat!" We were also armed with some medical books showing hideous sights of bones sticking through the skin, burns and bandaged accident cases, but hoped we would avoid such horrifics.

ENGINE

This was a 36 HP Isuzu C221 diesel, from Klassen Diesel, with a 2:1 reduction fluid transmission and a 17.5" x 11" folding propeller which gave us 5–6 knots. This was one of the few engines that had a pump attached by which one could simply accomplish an oil change in about ten minutes; a brilliant system.

Glossary

Anadromous: fish born and spawn in fresh water, spends life at sea.

Binnacle: box in front of helm holding compass

Bollard: a short vertical post for attaching mooring lines

Bootline: painted line on a hull close to and parallel to waterline

Burgee: a small identification flag, generally of a yacht club

Dodger: protection cover at the forward end of cockpit

Douse: lower the sail(s) completely

Fathom: 6 linear feet

Forepeak: interior area right in the bow

Genny: Genoa – foresail

Head: toilet

Knot: 1 nautical mile/hr or 1.85 kms/hr or 1.15 statute miles/hr

Painter: line or rope attached to dinghy

Pulpit: steel frame work in stern for general support

Glossary

Pushpit: steel framework in bow for general support

Reach: course sailed at more or less right angles to the wind

Skeg: the fixed forward end of a rudder system

Tender: dinghy for getting ashore

Yawing: boat twisting on and off about a vertical axis

Y.C.: yacht club